Elements of
Parallel Computing

Chapman & Hall/CRC
Computational Science Series

SERIES EDITOR

Horst Simon
Deputy Director
Lawrence Berkeley National Laboratory
Berkeley, California, U.S.A.

PUBLISHED TITLES

Elements of
Parallel Computing

Eric Aubanel

CRC Press
Taylor & Francis Group
Boca Raton London New York

CRC Press is an imprint of the
Taylor & Francis Group, an **informa** business

A CHAPMAN & HALL BOOK

CRC Press
Taylor & Francis Group
6000 Broken Sound Parkway NW, Suite 300
Boca Raton, FL 33487-2742

© 2017 by Taylor & Francis Group, LLC
CRC Press is an imprint of Taylor & Francis Group, an Informa business

No claim to original U.S. Government works

Printed on acid-free paper
Version Date: 20161028

International Standard Book Number-13: 978-1-4987-2789-1 (Paperback)

Visit the Taylor & Francis Web site at
http://www.taylorandfrancis.com

and the CRC Press Web site at
http://www.crcpress.com

To my grandfather, Dr. E.P. Scarlett: physician, educator and scholar.

Contents

Preface

Parallel computing is hard, it's creative, and it's an essential part of high performance scientific computing. I got my start in this field parallelizing quantum mechanical wave packet evolution for the IBM SP. Parallel computing has now joined the mainstream, thanks to multicore and manycore processors, and to the cloud and its Big Data applications. This ubiquity has resulted in a move to include parallel computing concepts in undergraduate computer science curricula. Clearly, a CS graduate must be familiar with the basic concepts and pitfalls of parallel computing, even if he/she only ever uses high level frameworks. After all, we expect graduates to have some knowledge of computer architecture, even if they never write code in an assembler.

Exposing undergraduates to parallel computing concepts doesn't mean dismantling the teaching of this subject in dedicated courses, as it remains an important discipline in computer science. I've found it a very challenging subject to teach effectively, for several reasons. One reason is that it requires students to have a strong background in sequential programming and algorithm design. Students with a shaky mental model of programming quickly get bogged down with parallel programming. Parallel computing courses attract many students, but many of them struggle with the challenges of parallel programming, debugging, and getting even modest speedup. Another challenge is that the discipline has been driven throughout its history by advances in hardware, and these advances keep coming at an impressive pace. I've regularly had to redesign my courses to keep up. Unfortunately, I've had little help from textbooks, as they have gone out of print or out of date.

This book presents the fundamental concepts of parallel computing not from the point of view of hardware, but from a more abstract view of the algorithmic and implementation patterns. While the hardware keeps changing, the same basic conceptual building blocks are reused. For instance, SIMD computation has survived through many incarnations from processor arrays to pipelined vector processors to SIMT execution on GPUs. Books on the theory of parallel computation approach the subject from a similar level of abstraction, but practical parallel programming books tend to be tied to particular programming models and hardware. I've been inspired by the work on parallel programming patterns, but I haven't adopted the formal design patterns approach, as I feel it is more suited to expert programmers than to novices.

My aim is to facilitate the teaching of parallel programming by surveying some key algorithmic structures and programming models, together with an abstract representation of the underlying hardware. The presentation is meant to be friendly and informal. The motivation, goals, and means of my approach are the subject of Chapter 1. The content of the book is language neutral, using pseudocode that represents common programming language models.

The first five chapters present the core concepts. After the introduction in Chapter 1, Chapter 2 presents SIMD, shared memory, and distributed memory machine models, along with a brief discussion of what their execution models look like. Chapter 2 concludes with a presentation of the task graph execution model that will be used in the following chapters. Chapter 3 discusses decomposition as a fundamental activity in parallel algorithmic design, starting with a naive example, and continuing with a discussion of some key algorithmic structures. Chapter 4 covers some important programming models in depth, and shows contrasting implementations of the task graphs presented in the previous chapter. Finally, Chapter 5 presents the important concepts of performance analysis, including work-depth analysis of task graphs, communication analysis of distributed memory algorithms, some key performance metrics, and a discussion of barriers to obtaining good performance. A brief discussion of how to measure performance and report performance is included because I have observed that this is often done poorly by students and even in the literature.

This book is meant for an introductory parallel computing course at the advanced undergraduate or beginning graduate level. A basic background in computer architecture and algorithm design and implementation is assumed. Clearly, hands-on experience with parallel programming is essential, and the instructor will have selected one or more languages and computing platforms. There are many good online and print resources for learning particular parallel programming language models, which could supplement the concepts presented in this book. While I think it's valuable for students to be familiar with all the parallel program structures in Chapter 4, in a given course a few could be studied in more detail along with practical programming experience. The instructor who wants to get students programming as soon as possible may want to supplement the algorithmic structures of Chapter 3 with simple programming examples. I have postponed performance analysis until Chapter 5, but sections of this chapter could be covered earlier. For instance, the work-depth analysis could be presented together with Chapter 3, and performance analysis and metrics could be presented in appropriate places in Chapter 4. The advice of Section 5.4 on reporting performance should be presented before students have to write up their experiments.

The second part of the book presents three case studies that reinforce the concepts of the earlier chapters. One feature of these chapters is to contrast different solutions to the same problem. I have tried for the most part to select problems that aren't discussed all that often in parallel computing textbooks. They include the Single Source Shortest Path Problem in Chapter 6, the Eikonal equation in Chapter 7, which is a partial differential equation with relevance to many fields, including graphics and AI, and finally in Chapter 8 a classical computational geometry problem, computation of the two-dimensional convex hull. These chapters could be supplemented with material from other sources on other well-known problems, such as dense matrix operations and the Fast Fourier Transform. I've also found it valuable, particularly in a graduate course, to have students research and present case studies from the literature.

Acknowledgements

I would first like to acknowledge an important mentor at the University of New Brunswick, professor Virendra Bhavsar, who helped me transition from a high performance computing practitioner to a researcher and teacher. I've used the generalized fractals he worked on, instead of the more common Mandelbrot set, to illustrate the need for load balancing. This book wouldn't have been possible without my experience teaching CS4745/6025, and the contributions of the students. In particular, my attention was drawn to the subset sum problem by Steven Stewart's course project and Master's thesis.

The formal panel sessions and informal discussions I witnessed from 2002 to 2014 at the International Parallel and Distributed Processing Symposium were very stimulating and influential in developing my views. I finally made the decision to write this book after reading the July 2014 JPDC special issue on Perspectives on Parallel and Distributed Processing. In this issue, Robert Schreiber's *A few bad ideas on the way to the triumph of parallel computing* [57] echoed many of my views, and inspired me to set them down in print. Finally, I would like to thank Siew Yin Chan, my former PhD student, for valuable discussions and revision of early material.

CHAPTER 1

Overview of Parallel Computing

1.1 INTRODUCTION

In the first 60 years of the electronic computer, beginning in 1940, computing performance per dollar increased on average by 55% per year [52]. This staggering 100 billion-fold increase hit a wall in the middle of the first decade of this century. The so-called *power wall* arose when processors couldn't work any faster because they couldn't dissipate the heat they produced. Performance has kept increasing since then, but only by placing multiple processors on the same chip, and limiting clock rates to a few GHz. These *multicore* processors are found in devices ranging from smartphones to servers.

Before the multicore revolution, programmers could rely on a free performance increase with each processor generation. However, the disparity between theoretical and achievable performance kept increasing, because processing speed grew much faster than memory bandwidth. Attaining peak performance required careful attention to memory access patterns in order to maximize re-use of data in cache memory. The multicore revolution made things much worse for programmers. Now increasing the performance of an application required parallel execution on multiple cores.

Enabling parallel execution on a few cores isn't too challenging, with support available from language extensions, compilers and runtime systems. The number of cores keeps increasing, and *manycore* processors, such as Graphics Processing Units (GPUs), can have thousands of cores. This makes achieving good performance more challenging, and parallel programming is required to exploit the potential of these parallel processors.

Why Learn Parallel Computing?

Compilers already exploit instruction level parallelism to speed up sequential code, so couldn't they also automatically generate multithreaded code to take advantage of multiple cores? Why learn distributed parallel programming, when frameworks such as MapReduce can meet the application programmer's needs? Unfortunately it's not so easy. Compilers can only try to optimize the code that's given to them, but they can't rewrite the underlying algorithms to be parallel. Frameworks, on the other hand, do offer significant benefits. However, they tend to be restricted to particular domains and they don't always produce the desired performance.

High level tools are very important, but we will always need to go deeper and apply parallel programming expertise to understand the performance of frameworks in order to make better use of them. Specialized parallel code is often essential for applications requiring high performance. The challenge posed by the rapidly growing number of cores has meant that more programmers than ever need to understand something about parallel programming. Fortunately parallel processing is natural for humans, as our brains have been described as parallel processors, even though we have been taught to program in a sequential manner.

Why is a New Approach Needed?

Rapid advances in hardware make parallel computing exciting for the devotee. Unfortunately, advances in the field tend to be driven by the hardware. This has resulted in solutions tied to particular architectures that quickly go out of date, as do textbooks. On the software front it's easy to become lost amid the large number of parallel programming languages and environments.

Good parallel algorithm design requires postponing consideration of the hardware, but at the same time good algorithms and implementations must take the hardware into account. The way out of this parallel software/hardware thicket is to find a suitable level of abstraction and to recognize that a limited number of solutions keep being re-used.

This book will guide you toward being able to think in parallel, using a task graph model for parallel computation. It makes use of recent work on parallel programming patterns to identify commonly used algorithmic and implementation strategies. It takes a language-neutral approach using pseudocode that reflects commonly used language models. The pseudocode can quite easily be adapted for implementation in relevant languages, and there are many good resources online and in print for parallel languages.

1.2 TERMINOLOGY

It's important to be clear about terminology, since *parallel computing*, *distributed computing*, and *concurrency* are all overlapping concepts that have been defined in different ways. Parallel computers can also be placed in several categories.

Definition 1.1 (Parallel Computing). *Parallel Computing means solving a computing problem in less time by breaking it down into parts and computing those parts simultaneously.*

Parallel computers provide more computing resources and memory in order to tackle problems that cannot be solved in a reasonable time by a single processor core. They differ from sequential computers in that there are multiple processing elements that can execute instructions in parallel, as directed by the parallel program. We can think of a sequential

Instruction streams

		single	multiple
Data streams	single	SISD	MISD
	multiple	SIMD	MIMD

Figure 1.1: Flynn's Taxonomy.

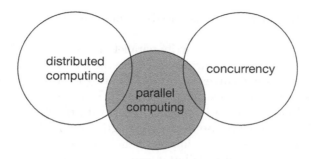

Figure 1.2: Three overlapping disciplines.

computer, as described by the von Neumann architecture, as executing a stream of instructions that accesses a stream of data. Parallel computers work with multiple streams of data and/or instructions, which is the basis of Flynn's taxonomy, given in Figure 1.1. A sequential computer is in the *Single Instruction Single Data* (SISD) category and parallel computers are in the other categories. *Single Instruction Multiple Data* (SIMD) computers have a single stream of instructions that operate on multiple streams of data in parallel. *Multiple Instruction Multiple Data* (MIMD) is the most general category, where each instruction stream can operate on different data. There aren't currently any *Multiple Instruction Single Data* (MISD) computers in production. While most parallel computers are in the MIMD category, most also incorporate SIMD processing elements.

MIMD computers are further classified into *shared memory* and *distributed memory* computers. The processing elements of a shared memory computer all have access to a single memory address space. Distributed memory computers have memory that is distributed among compute nodes. If a processing element on one node wants to access data on another node, it can't directly refer to it by its address, but must obtain it from the other node by exchanging messages.

Distributed memory parallel computing can be done on a cluster of computers connected by a network. The larger field of distributed computing has some of the same concerns, such as speed, but is not mainly concerned with the solution of a single problem. Distributed systems are typically loosely coupled, such as peer-to-peer systems, and the main concerns include reliability as well as performance.

Parallel computing can also be done on a shared memory multiprocessor. Simultaneous access to the same data can lead to incorrect results. Techniques from the field of concurrency, such as mutual exclusion, can be used to ensure correctness. The discipline of concurrency is about much more than parallel computing. Shared access to data is also important for databases and operating systems. For example, concurrency addresses the problem of ensuring that two simultaneous operations on a bank account don't conflict.

We can summarize the three overlapping disciplines with the diagram in Figure 1.2.

1.3 EVOLUTION OF PARALLEL COMPUTERS

Parallel computers used to be solely large expensive machines with specialized hardware. They were housed at educational and governmental institutions and were mainly dedicated to scientific computing. An important development came in the 1990s with the so-called Beowulf revolution, where the best performance to cost ratio could be obtained with clusters of commodity PCs connected by network switches rather than with expensive purpose-built supercomputers. While this made parallel computing more accessible, it still remained the domain of enthusiasts and those whose needs required high performance computing resources.

Not only is expert knowledge needed to write parallel programs for a cluster, the significant computing needs of large scale applications require the use of shared supercomputing facilities. Users had to master the complexities of coordinating the execution of applications and file management, sometimes across different administrative and geographic domains. This led to the idea of Grid computing in the late 1990s, with the dream of computing as a utility, which would be as simple to use as the power grid. While the dream hasn't been realized, Grid computing has provided significant benefit to collaborative scientific data analysis and simulation. This type of distributed computing wasn't adopted by the wider community until the development of cloud computing in the early 2000s.

Cloud computing has grown rapidly, thanks to improved network access and virtualization techniques. It allows users to rent computing resources on demand. While using the cloud doesn't require parallel programming, it does remove financial barriers to the use of compute clusters, as these can be assembled and configured on demand. The introduction of frameworks based on the MapReduce programming model eliminated the difficulty of parallel programming for a large class of data processing applications, particularly those associated with the mining of large volumes of data.

With the emergence of multicore and manycore processors all computers are parallel computers. A desktop computer with an attached manycore co-processor features thousands of cores and offers performance in the trillions of operations per second. Put a number of these computers on a network and even more performance is available, with the main limiting factor being power consumption and heat dissipation. Parallel computing is now relevant to all application areas. Scientific computing isn't the only player any more in large scale high performance computing. The need to make sense of the vast quantity of data that cheap computing and networks have produced, so-called Big Data, has created another important use for parallel computing.

1.4 EXAMPLE: WORD COUNT

Let's consider a simple problem that can be solved using parallel computing. We wish to list all the words in a collection of documents, together with their frequency. We can use a list containing key-value pairs to store words and their frequency. The sequential Algorithm 1.1 is straightforward.

Algorithm 1.1: Sequential word count

Input: collection of text documents
Output: list of $\langle word, count \rangle$ pairs

foreach *document in collection* **do**
    foreach word *in document* **do**
        if *first occurrence of* word **then**
            add $\langle word, 1 \rangle$ to ordered list
        else
            increment count in $\langle word, count \rangle$
        end
    end
end

If the input consists of two documents, one containing "The quick brown fox jumps over a lazy dog" and the other containing "The brown dog chases the tabby cat," then the output would be the list: $\langle a, 1 \rangle, \langle brown, 2 \rangle, \ldots, \langle tabby, 1 \rangle, \langle the, 3 \rangle$.

Think about how you would break this algorithm in parts that could be computed in parallel, before we discuss below how this could be done.

1.5 PARALLEL PROGRAMMING MODELS

Many programming models have been proposed over the more than 40 year history of parallel computing. Parallel computing involves identifying those tasks that can be performed concurrently and coordinating their execution. Fortunately, abstraction is our friend, as in other areas of computer science. It can mask some of the complexities involved in implementing algorithms that make use of parallel execution. Programming models that enable parallel execution operate at different levels of abstraction. While only a few will be mentioned here, parallel programming models can be divided into three categories, where the parallelism is implicit, partly explicit, and completely explicit [66].

1.5.1 Implicit Models

There are some languages where parallelism is implicit. This is the case for functional programming languages, such as Haskell, where the runtime works with the program as a graph and can identify those functions that can be executed concurrently. There are also algorithmic skeletons, which may be separate languages or frameworks built on existing languages. Applications are composed of parallel skeletons, such as in the case of MapReduce. The programmer is concerned with the functionality of the components of the skeletons, not with how they will be executed in parallel.

The word count problem solved in Algorithm 1.1 is a canonical MapReduce application. The programmer writes a `map` function that emits ⟨word, count⟩ pairs and a `reduce` function to sum the values of all pairs with the same word. We'll examine MapReduce in more detail below and in Chapter 4.

1.5.2 Semi-Implicit Models

There are other languages where programmers identify regions of code that can be executed concurrently, but where they do not have to determine how many resources (threads/processes) to use and how to assign tasks to them. The Cilk language, based on C/C++, allows the programmer to identify recursive calls that can be done in parallel and to annotate loops whose iterations can be computed independently. OpenMP is another approach, which augments existing imperative languages with an API to enable loop and task level specification of parallelism. Java supports parallelism in several ways. The Fork/Join framework of Java 7 provides an Executor service that supports recursive parallelism much in the same way as Cilk. The introduction of streams in Java 8 makes it possible to identify streams that can be decomposed into parallel sub-streams. For these languages the compiler and runtime systems take care of the assignment of tasks to threads. Semi-implicit models are becoming more relevant with the increasing size and complexity of parallel computers.

For example, parallel loops can be identified by the programmer, such as:

parallel for $i \leftarrow 0$ *to* $n - 1$ **do**
  $c[i] \leftarrow a[i] + b[i]$
end

Once loops that have independent iterations have been identified, the parallelism is very simple to express. Ensuring correctness is another matter, as we'll see in Chapters 2 and 4.

1.5.3 Explicit Models

Finally, there are lower level programming models where parallelism is completely explicit. The most popular has been the Message Passing Interface (MPI), a library that enables parallel programming for C/C++ and Fortran. Here the programmer is responsible for identifying tasks, mapping them to processors, and sending messages between processors. MPI has been very successful due to its proven performance and portability, and the ease with which parallel algorithms can be expressed. However, MPI programs can require significant development time, and achieving the portability of performance across platforms can be very challenging. OpenMP, mentioned above, can be used in a similar way, where all parallelism and mapping to threads is expressed explicitly. It has the advantage over MPI of a shared address space, but can suffer from poorer performance, and is restricted to platforms that offer shared memory. So-called partitioned global address space (PGAS) languages combine some of the advantages of MPI and OpenMP. These languages support data parallelism by allowing the programmer to specify data structures that can be distributed across processes, while providing the familiar view of a single address space. Examples include Unified Parallel C and Chapel.

In the following message passing example for the sum of two arrays a and b on p processors, where the arrays are initially on processor 0, the programmer has to explicitly scatter the operands and gather the results:

scatter(0, a, n/p, $aLoc$)
scatter(0, b, n/p, $bLoc$)
for $i \leftarrow 0$ *to* $n/p - 1$ **do**
    $cLoc[i] \leftarrow aLoc[i] + bLoc[i]$
end
gather(0, c, n/p, $cLoc$)

Here arrays a and b of length n are scattered in contiguous chunks of n/p elements to p processors, and stored in arrays $aLoc$ and $bLoc$ on each processor. The resulting $cLoc$ arrays are gathered into the c array on processor 0. We'll examine message passing in detail in Chapter 4.

1.5.4 Thinking in Parallel

Parallel programming models where the parallelism is implicit don't require programmers to "think in parallel." As long as the underlying model (e.g., functional language or skeleton framework) is familiar, then the benefit of harnessing multiple cores can be achieved without the additional development time required by explicit parallel programming. However, not only can this approach limit the attainable performance, more importantly it limits the exploration space in the design of algorithms. Learning to think in parallel exposes a broader landscape of algorithm design. The difficulty is that it requires a shift in the mental model of the algorithm designer or programmer. Skilled programmers can look at code and run it in their mind, executing it on the *notional machine* associated with the programming language. A notional machine explains how a programming language's statements are executed. Being able to view a static program as something that is dynamically executed on a notional machine is a *threshold concept*. Threshold concepts are transformative and lead to new

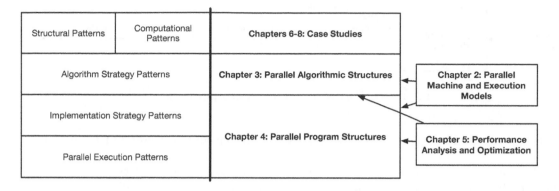

Figure 1.3: OPL hierarchy (left) and corresponding chapters (right).

ways of thinking [69]. It could be argued that making the transition from sequential to parallel notional machines is another threshold concept.

1.6 PARALLEL DESIGN PATTERNS

While the prospect of an enlarged algorithmic design space may be thrilling to some, it can be daunting to many. Fortunately, we can build on the work of the parallel programming community by reusing existing algorithmic techniques. This has been successful particularly in object-oriented software design, where design patterns offer guidance to common software design problems at different layers of abstraction. This idea has been extended to parallel programming, notably with Berkeley's Our Pattern Language (OPL) [44]. Programmers can identify patterns in program structure at a high level of abstraction, such as pipe-and-filter, or those that are found in particular application domains, such as in graph algorithms.

Although these high level patterns do not mention parallelism, they can naturally suggest parallel implementation, as in the case of pipe-and-filter, which can benefit from pipelined parallel execution. These patterns also document which lower level patterns are appropriate. At lower levels there are patterns that are commonly used in algorithm and software design, such as divide-and-conquer, geometric decomposition, and the master-worker pattern. There is a seemingly never ending number of algorithmic and design patterns, which is why mastering the discipline can take a long time. The true benefit of the work of classifying patterns is that it can provide a map of parallel computing techniques.

Figure 1.3 illustrates the OPL hierarchy and indicates how the chapters of this book refer to different layers. The top two layers are discussed in this chapter. We will not be considering the formal design patterns in the lower layers, but will examine in Chapters 3 and 4 the algorithmic and implementation structures that they cover.

1.6.1 Structural Patterns

Structural patterns consist of interacting components that describe the high level structure of a software application. These include well known structures that have been studied by the design pattern community. Examples include the model-view-controller pattern that is used in graphical user interface frameworks, the pipe-and-filter pattern for applying a series of filters to a stream of data, and the agent-and-repository and MapReduce patterns used

Figure 1.4: Part of a text processing pipeline.

in data analysis. These patterns can be represented graphically as a group of interacting tasks.

Consider the pipe-and-filter pattern, which is useful when a series of transformations needs to be applied to a collection of data. The transformations, also called filters, can be organized in a pipeline. Each filter is independent and does not produce side effects. It reads its input, applies its transformation, then outputs the result. As in a factory assembly line, once the stream of data fills the pipeline the filters can be executed in parallel on different processors. If the filters take roughly the same amount of time, the execution time can be reduced proportionately to the number of processors used. It can happen that some filters will take much longer to execute than others, which can cause a bottleneck and slow down execution. The slower filters can be sped up by using parallel processing techniques to harness multiple processors. An example is shown in Figure 1.4, which shows the first few stages of a text processing pipeline. A stream of social media messages is first filtered to only keep messages in English, then any metadata (such as URLs) is removed in the next filter. The third filter tokenizes the messages into words. The pipeline could continue with other filters to perform operations such as tagging and classification.

Another pattern, MapReduce, became popular when Google introduced the framework of the same name in 2004, but it has been used for much longer. It consists of a map phase, where the same operation is performed on objects in a collection, followed by a reduce phase where a summary of the results of the map phase is collected. Many applications that need to process and summarize large quantities of data fit this pattern. Continuing with the text processing example, we might want to produce a list of words and their frequency from a stream of social media messages, as in the example of Section 1.4.

Both map and reduce phases of this pattern can be executed in parallel. Since the map operations are independent, they can easily be done in parallel across multiple processors. This type of parallel execution is sometimes called embarrassingly parallel, since there are no dependencies between the tasks and they can be trivially executed in parallel. The reduce phase can be executed in parallel using the reduction operation, which is a frequently used lower level parallel execution pattern. The attraction of MapReduce, implemented in a framework, is that the developer can build the mapper and reducer without any knowledge of how they are executed.

1.6.2 Computational Patterns

Whereas in cases like pipe-and-filter and MapReduce the structure reveals the potential for parallel execution, usually parallelism is found in the functions that make up the components of the software architecture. In practice these functions are usually constructed from a limited set of computational patterns. Dense and sparse linear algebra computations are probably the two most common patterns. They are used by applications such as games, machine learning, image processing and high performance scientific computing. There are well established parallel algorithms for these patterns, which have been implemented in many

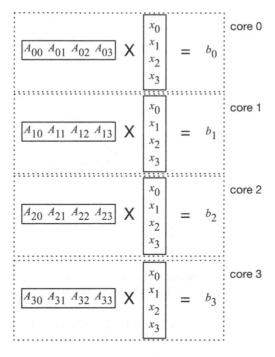

Figure 1.5: Parallel matrix-vector multiplication $b = Ax$.

libraries. The High Performance Linpack (HPL) benchmark used to classify the top 500 computers involves solving a dense system of linear equations. Parallelism arises naturally in these applications. In matrix-vector multiplication, for example, the inner products that compute each element of the result vector can be computed independently, and hence in parallel, as seen in Figure 1.5. In practice it is more difficult to develop solutions that scale well with matrix size and the number of processors, but the plentiful literature provides guidance.

Another important pattern is one where operations are performed on a grid of data. It occurs in scientific simulations that numerically solve partial differential equations, and also in image processing that executes operations on pixels. The solutions for each data point can be computed independently but they require data from neighboring points. Other patterns include those found in graph algorithms, optimization (backtrack/branch and bound, dynamic programming), and sorting. It is reassuring that the lists that have been drawn up of these patterns include less than twenty patterns. Even though the landscape of parallel computing is vast, most applications can be composed of a small number of well studied computational patterns.

1.6.3 Patterns in the Lower Layers

Structural and computational patterns allow developers to identify opportunities for parallelism and exploit ready-made solutions by using frameworks and libraries, without the need to acquire expertise in parallel programming. Exploration of the patterns in the lower layers is necessary for those who wish to develop new parallel frameworks and libraries, or need a customized solution that will obtain higher performance. This requires the use of

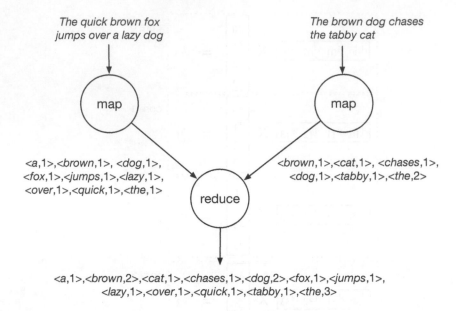

Figure 1.6: Word count as a MapReduce pattern.

algorithmic and implementation structures to create parallel software. These structures will be explored in detail in Chapters 3 and 4.

1.7 WORD COUNT IN PARALLEL

Let's return to the word count problem of Section 1.4. Figure 1.6 illustrates how this algorithm can be viewed as an instance of the MapReduce pattern. In the map phase, ⟨word, count⟩ pairs are produced for each document. The reduce phase aggregates the pairs produced from all documents. What's not shown in Figure 1.6 is that there can be multiple reducers. If we're using a MapReduce framework then we just need to implement a mapper function to generate ⟨word, count⟩ pairs given a document and a reducer function to sum the values of a given word, as we'll see in Chapter 4.

To produce an explicitly parallel solution we need to start by finding the parallelism in the problem. The task of creating ⟨word, count⟩ pairs can be done independently on each document, and therefore can be done trivially in parallel. The reduction phase is not as simple, since some coordination among tasks is needed, as described in the reduction algorithm pattern. Next, we need to consider what type of computational platform is to be used, and whether the documents are stored locally or are distributed over multiple computers. A distributed approach is required if the documents are not stored in one place or if the number of documents is large enough to justify using a cluster of computers, which might be located in the cloud.

Alternatively, a local computer offers the choice of using a shared data structure. A concurrent hash map would allow multiple threads to update ⟨word, count⟩ pairs, while providing the necessary synchronization to avoid conflicts should multiple updates overlap. A distributed implementation could use the master-worker pattern, where the master performs the distribution and collection of work among the workers. Note that this approach could also be used on multiple cores of a single computer.

Both distributed and shared implementations could be combined. Taking a look at the sequential algorithm we can see that word counts could be done independently for each line of text. The processing of the words of the sentences of each document could be accomplished in parallel using a shared data structure, while the documents could be processed by multiple computers.

This is a glimpse of how rich the possibilities can be in parallel programming. It can get even richer as new computational platforms emerge, as happened in the mid 2000s with general purpose programming on graphics processing units (GPGPU). The patterns of parallel computing provided a guide in our example problem, as they allowed the recognition that it fit the MapReduce structural pattern, and that implementation could be accomplished using well established algorithmic and implementation patterns. While this book will not follow the formalism of design patterns, it does adopt the similar view that there is a set of parallel computing elements that can be composed in many ways to produce clear and effective solutions to computationally demanding problems.

1.8   OUTLINE OF THE BOOK

Parallel programming requires some knowledge of parallel computer organization. Chapter 2 begins with a discussion of three abstract machine models: SIMD, shared memory, and distributed memory. It also discusses the hazards of access to shared variables, which threaten program correctness. Chapter 2 then presents the task graph as an execution model for parallel computing. This model is used for algorithm design, analysis, and implementation in the following chapters. Chapter 3 presents an overview of algorithmic structures that are often used as building blocks. It focuses on the fundamental task of finding parallelism via task and data decomposition. Chapter 4 explores the three machine models more deeply through examination of the implementation structures that are relevant to each model.

Performance is a central concern for parallel programmers. Chapter 5 first shows how work-depth analysis allows evaluation of the task graph of a parallel algorithm, without considering the target machine model. The barriers to achieving good performance that are encountered when implementing parallel algorithms are discussed next. This chapter closes with advice on how to effectively and honestly present performance results.

The final chapters contain three detailed case studies, drawing on what was learned in the previous chapters. They consider different types of parallel solutions for different machine models, and take a step-by-step voyage through the algorithm design and analysis process.

CHAPTER 2

Parallel Machine and Execution Models

Unsurprisingly, parallel programs are executed on parallel machines. It may be less obvious that they are also executed in the mind of the parallel programmer. All programmers have a mental model of program execution that allows them to trace through their code. Programmers who care about performance also have a mental model of the machine they are using. Knowledge about the operation of the cache hierarchy, for instance, allows a programmer to structure loops to promote cache reuse. While many programmers can productively work at a high level of abstraction without any concern for the hardware, parallel programmers cannot afford to do so. The first part of this chapter discusses three machine models and the second part presents a task graph execution model. The task graph model is used extensively in Chapter 3 when presenting algorithmic structures. The machine models are examined more deeply together with implementation structures in Chapter 4.

2.1 PARALLEL MACHINE MODELS

We will not delve into the details of actual parallel computers, because they continue to evolve at a very rapid pace and an abstract representation is more appropriate for our purposes. We'll look instead at some general features that can be observed when surveying parallel computers past and present. Stated simply, parallel computers have multiple processing cores. They range from the most basic multi-core processor to clusters of hundreds of thousands of processors. They also vary in the functionality of the cores and in how they are interconnected.

Modern microprocessors have become incredibly complicated, as they have used the ever increasing number of transistors to improve the performance of execution of a single instruction stream, using techniques such as pipelining and superscalar execution. They have also devoted an increasing fraction of the chip to high speed cache memory, in an attempt to compensate for slow access to main memory. Because of this, multicore processors have a limited number of cores. So-called *manycore* processors simplify the design of the cores so that more can be placed on a chip, and therefore more execution streams can proceed in parallel. These two types of architectures have been called *latency oriented* and *throughput oriented*. They can be seen in general purpose microprocessors on one hand, and graphics processing units (GPUs) on the other [28].

Three machine models can represent, individually and in combination, all current gen-

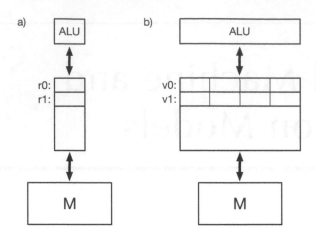

Figure 2.1: a) scalar architecture b) SIMD architecture, with vector registers than can contain 4 data elements.

eral purpose parallel computers: *SIMD*, *shared memory* and *distributed memory*. Recall from Chapter 1 that the latter two are subcategories of the MIMD architecture. These machines mainly need to be understood at a high level of abstraction for the purpose of algorithm design and implementation strategy. Understanding of deeper levels of machine architecture is important during implementation, at which point technical documentation from the hardware vendor should be consulted. One exception has to do with access to shared variables, where race conditions can lead to incorrect programs. Another is the performance impact of cache memory, which is sensitive to data access patterns.

2.1.1 SIMD

SIMD computers simultaneously execute a single instruction on multiple data elements. Historically, processors in SIMD computers consisted either of a network of Arithmetic Logic Units (ALUs), as in the Connection Machine, or had deeply pipelined vector arithmetic units, as in computers from Cray and NEC. Currently, SIMD execution is mainly found in functional units in general purpose processors, and our discussion will reflect this architecture, as sketched in Figure 2.1. SIMD ALUs are wider than conventional ALUs and can perform multiple operations simultaneously in a single clock cycle. They use wide SIMD registers that can load and store multiple data elements to memory in a single transaction.

For example, for a scalar ALU to add the first 4 elements of two arrays a and b and store the result in array c, machine instructions similar to the following would need to be executed in sequence:

1: $r1 \leftarrow$ load $a[0]$	**5:** $r1 \leftarrow$ load $a[1]$	**9:** $r1 \leftarrow$ load $a[2]$	**13:** $r1 \leftarrow$ load $a[3]$
2: $r2 \leftarrow$ load $b[0]$	**6:** $r2 \leftarrow$ load $b[1]$	**10:** $r2 \leftarrow$ load $b[2]$	**14:** $r2 \leftarrow$ load $b[3]$
3: $r2 \leftarrow$ add $r1, r2$	**7:** $r2 \leftarrow$ add $r1, r2$	**11:** $r2 \leftarrow$ add $r1, r2$	**15:** $r2 \leftarrow$ add $r1, r2$
4: $c[0] \leftarrow$ store $r2$	**8:** $c[1] \leftarrow$ store $r2$	**12:** $c[2] \leftarrow$ store $r2$	**16:** $c[3] \leftarrow$ store $r2$

A SIMD ALU of width equal to the size of four elements of the arrays could do the same operations in just four instructions:

1: $v1 \leftarrow$ vload a
2: $v2 \leftarrow$ vload b
3: $v2 \leftarrow$ vadd $v1, v2$
4: $c \leftarrow$ vstore $v2$

The *vload* and *vstore* instructions load and store four elements of an array from memory into a vector register in a single transaction. The *vadd* instruction simultaneously adds the four values stored in each vector register.

SIMD execution is also found in manycore co-processors, such as the Intel Xeon Phi. Nvidia Graphics Processing Units (GPUs) use SIMD in a different way, by scheduling threads in groups (called *warps*), where the threads in each group perform the same instruction simultaneously on their own data. The difference with conventional SIMD led the company to coin a new term, *Single Instruction Multiple Threads* (SIMT). It works something like this:

t0: $r00 \leftarrow$ load $a[0]$	$r10 \leftarrow$ load $a[1]$	$r20 \leftarrow$ load $a[2]$	$r30 \leftarrow$ load $a[3]$
t1: $r01 \leftarrow$ load $b[0]$	$r11 \leftarrow$ load $b[1]$	$r21 \leftarrow$ load $b[2]$	$r31 \leftarrow$ load $b[3]$
t2: $r01 \leftarrow$ add $r00, r01$	$r11 \leftarrow$ add $r10, r11$	$r21 \leftarrow$ add $r20, r21$	$r31 \leftarrow$ add $r30, r31$
t3: $c[0] \leftarrow$ store $r01$	$c[1] \leftarrow$ store $r11$	$c[2] \leftarrow$ store $r21$	$c[3] \leftarrow$ store $r31$

Thread i loads $a[i]$, and all loads are coalesced into a single transaction at time **t0**. The same thing occurs at time **t1** for loading the elements of b. All threads simultaneously perform an addition at time **t2**. Finally all stores to c are coalesced at time **t3**. The difference between SIMD and SIMT for the programmer is in the programming model: SIMD uses data parallel loops or array operations whereas in the SIMT model the programmer specifies work for each thread, and the threads are executed in SIMD fashion.

Connection Machine: Art and Science

Computers are becoming invisible as they become ubiquitous, either disappearing into devices or off in the cloud somewhere. Supercomputers tend to be physically impressive because of the space they occupy and the visible cables, cooling, and power infrastructure. But to the uninitiated they are not very impressive because their design doesn't express how they function. The Connection Machine was an exception. It looked like a work of art, because in a way it was. Produced by Thinking Machines between 1983 and 1994, it was the realization of Danny Hillis's PhD research at MIT. The Connection Machine was a massively parallel SIMD computer. The CM-1 had 4096 chips with 16 processors each, for a total of 65,536 processors. The chips were connected in a 12-dimensional hypercube. The task of designing the enclosure for the CM-1 was assigned to the artist Tamiko Thiel. With the help of Nobel winning physicist Richard Feynman, she designed the striking cubic case with transparent sides that revealed red LED lights indicating the activity of the processors. These blinking lights not only helped programmers verify that as many processors were occupied as possible, but they made the computer look like a living, thinking machine. The Connection Machine was originally designed to solve problems in artificial intelligence, and was programmed using a version of Lisp. During his brief stay at Thinking Machines, Feynman also demonstrated that the Connection Machine was good for scientific problems by writing a program to numerically solve problems in Quantum Chromodynamics (QCD). Feynman's QCD solver outperformed a computer that Caltech, his home institution, was building to solve QCD problems.

The popularity of the SIMD architecture comes from the common programming pattern

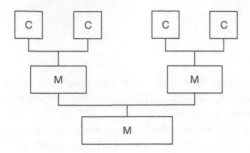

Figure 2.2: Basic multicore processor model, consisting of multiple cores (C) sharing memory (M) units organized hierarchically.

where the same operation is executed independently on the elements of an array. SIMD units can be programmed explicitly using a parallel programming language that supports data parallelism. They can also be programmed directly in assembly language using vector instructions, as in the above example. More commonly the parallelism is not specified explicitly by the programmer, but is exploited by the compiler or by the runtime system.

The SIMD model is not hard to understand from the programmer's perspective, but it can be challenging to work within the restrictions imposed by the model. Common execution patterns often deviate from the model, such as conditional execution and non-contiguous data, as we'll see in Chapter 4.

2.1.2 Shared Memory and Distributed Memory Computers

Multicore Processor

Most parallel computers fall into the Multiple Instruction Multiple Data (MIMD) category, as each execution unit (process or thread) can perform different instructions on different data. These computers differ in how the processing units are connected to each other and to memory modules. A notable feature that unites them is that the memory modules are organized hierarchically.

Figure 2.2 shows a model of a hypothetical multicore processor. The memory node at the root represents the main off-chip memory associated with the processor. When a core loads or stores a data element, a block of data containing that element is copied into its local cache, represented in Figure 2.2 as a child of the main memory node. Access to off-chip memory is typically several orders of magnitude slower than access to on-chip cache memory. While all cores have access to the entire memory address space, those sharing a cache are closer together in the sense that they can share data more efficiently among each other than with cores in the other group. The memory hierarchy is normally invisible to programmers, which means they do not have to be concerned with moving data between the memory units, although some processors (e.g. Nvidia GPUs) have memory local to a group of cores that is managed explicitly by the programmer.

Multiprocessors and Multicomputers

More processing power can be obtained by connecting several multicore processors together, as in Figure 2.3. This machine can be realized in two ways. First, it can provide a single memory address space to all processing units. This can be realized in a single *multiprocessor* computer, where an on-board network connects the memory modules of each processor.

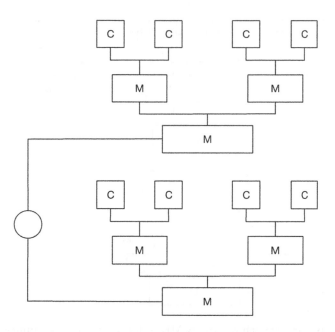

Figure 2.3: Basic parallel computer with two processors. Can be implemented as shared memory (multiprocessor) or distributed memory (multicomputer).

It can also be realized in a *distributed shared memory* computer as a network of multiprocessors, where a single address space is enabled either in hardware or software. These multiprocessors implement what is called a Non Uniform Memory Architecture (NUMA), due to the fact that memory access times are not uniform across the address space. A core in one processor will obtain data more quickly from its processor's memory than from another processor. This difference in access times can be as much as a factor of two or more. In contrast, the architecture in Figure 2.2 can be called UMA, since access times to an element of data will be the same for all cores, assuming it is not already in any of the caches.

Second, each processor may have a separate address space and therefore messages need to be sent between processors to obtain data not stored locally. This is called a *distributed multicomputer*. In either case it is clear that increasing the ratio of local to nonlocal memory accesses can provide a performance gain.

The multiprocessor in Figure 2.3 can be expanded by adding more processors at the same level, or combining several multiprocessors to form a deeper hierarchy. Knowledge of the hierarchical structure of the machine can guide program design, for instance by encouraging a hierarchical decomposition of tasks. The actual network topology is not relevant to the parallel programmer, however. It is possible to design the communication pattern of a parallel program tailored to a given network topology. This usually isn't practical as it limits the type of machine the program will perform well on. However, the performance characteristics of the network are relevant when implementing an algorithm. For instance, if point-to-point latency is high but there is plenty of bandwidth then it may be better to avoid sending many small messages and to send them in larger batches instead. We'll look at performance analysis of distributed memory programs in Chapter 5.

Figure 2.4 shows a machine model for a hybrid multicore/manycore processor. The manycore co-processor follows the shared memory model, as all cores have access to the same memory address space. It also incorporates SIMD execution, either through SIMD

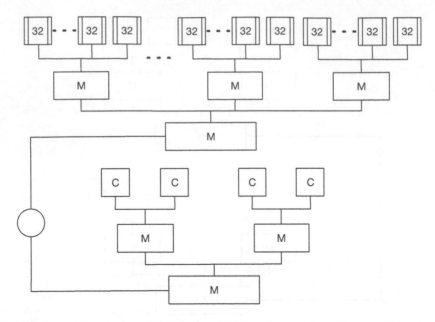

Figure 2.4: Manycore processor with 32-way SIMD units connected to a multicore processor

functional units, as in the Intel Xeon Phi, or SIMT execution of threads in a warp on an Nvidia GPU. The multicore processor and manycore co-processor can share memory, or they can have distinct memory spaces. In either case there are likely to be nonuniform access times to memory, as in the multiprocessor of Figure 2.3. Multiple hybrid multicore/manycore computers can be combined on a network to produce a distributed multicomputer with very high performance.

2.1.3 Distributed Memory Execution

Parallel execution in the distributed memory model involves execution of multiple processes, with each process normally assigned to its own processor core. Each process has its own private memory, and can only access data in another process if that process sends the data in a message. Distributed memory execution is suited to multicomputers, but it can also be applied to shared memory multiprocessors.

To continue with our simple array example, consider adding the four elements with two processes, this time with pseudocode representing a higher level language than assembler to keep the exposition brief:

```
   // process 0                          // process 1
1: send a[2...3] to process 1         1: receive a[0...1] from process 0
2: send b[2...3] to process 1         2: receive b[0...1] from process 0
3: c[0] ← a[0] + b[0]                 3: c[0] ← a[0] + b[0]
4: c[1] ← a[1] + b[1]                 4: c[1] ← a[1] + b[1]
5: receive c[2...3] from process 1    5: send c[0...1] to process 0
```

Arrays a and b are initially in the memory of process 0, and the result of $a+b$ is to be stored there as well. The communication is illustrated in Figure 2.5. Process 0 sends the second half of arrays a and b to process 1. Notice that each process uses local indexing for the

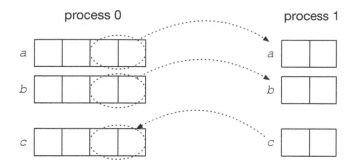

Figure 2.5: Communication involved in adding two arrays with two processes.

arrays, and hence both use identical statements to add their arrays. Execution of the sums can then take place independently on each process until process 1 has finished its sums and sends its array c to process 0. Lines 3 and 4 won't necessarily execute exactly simultaneously. Process 0 might start earlier since it could proceed to line 3 while both messages are in flight and before process 1 has received them. The execution times of these operations could be different for each process, due to different memory access times or competition for CPU resources from operating system processes.

The message passing programming model gives a lot of control to the programmer. Since communication is explicit it's possible to trace through execution and reason about correctness and performance. In practice this type of programming can be quite complex. We'll look at some common patterns in message passing programming in Chapter 4.

2.1.4 Shared Memory Execution

Threads are the execution units for shared memory programs. Unlike processes they share memory with other threads, so they do not need to use explicit communication to share data. They can follow their own execution path with private data stored on the runtime stack. Execution of our array sum by four threads would look like this:

```
// thread 0            // thread 1            // thread 2            // thread 3
r00 ← load a[0]        r10 ← load a[1]        r20 ← load a[2]        r30 ← load a[3]
r01 ← load b[0]        r11 ← load b[1]        r21 ← load b[2]        r31 ← load b[3]
r01 ← add r00,r01      r11 ← add r10,r11      r21 ← add r20,r21      r31 ← add r30,r31
c[0] ← store r01       c[1] ← store r11       c[2] ← store r21       c[3] ← store r31
```

Here the threads don't execute their instructions in lockstep, unlike the SIMT example above. Their execution will overlap in time, but the exact timing is unpredictable. Observe that each thread loads and stores to different memory locations, so the relative timing of threads has no impact on the result.

Data Races

Let's say after adding a and b we want to use array c, and we add two more instructions for thread 0:

```
r00 ← load c[3]
x ← store r00
```

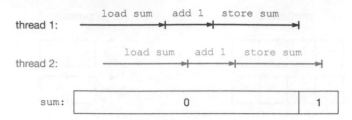

Figure 2.6: Example of a data race.

This means thread 0 is loading from a memory location that thread 3 stores to. The value of x is indeterminate, since it depends on the relative timing of threads 0 and 3. The problem is that thread 0 could load $c[3]$ before thread 3 stored its value to $c[3]$. This is an example of a *data race*, which occurs when one thread stores a value to a memory location while other threads load or store to the same location. Data races are a serious threat to program correctness.

Data races can be hidden from view in higher level programming languages. Consider two threads incrementing the same variable:

```
// thread 0          // thread 1
sum ← sum + 1         sum ← sum + 1
```

If *sum* is initially 0, then one might think that the result of these two operations would be $sum = 2$, whatever the temporal order of the two threads. However, another possible result is $sum = 1$, because the statement $sum \leftarrow sum + 1$ is not executed *atomically* (from the Greek *atomos* meaning indivisible) at the machine level. Let's look at machine instructions for this operation:

```
// thread 0          // thread 1
r00 ← load sum        r10 ← load sum
r01 ← add r00,1       r11 ← add r10,1
sum ← store r01       sum ← store r11
```

Now the data race is more easily seen, as both threads could load *sum* before either of them stored their value, as illustrated in Figure 2.6.

Returning to our previous example, we could try to fix this data race by synchronizing threads so that thread 0 doesn't read from $c[3]$ until thread 3 has written its value there. Thread 3 sets a flag once its value is written, and thread 0 makes sure that flag is set before loading $c[3]$ (flag is initialized to 0):

```
// thread 0              // thread 3
L: r00 ← load flag       c[3] ← store r31
If flag ≠ 1 goto L       flag ← store 1
r00 ← load c[3]
x ← store r00
```

We might expect that, even if thread 3 were delayed, thread 0 would keep looping until thread 3 had written to $c[3]$ and set the flag. This expectation comes from our practice of reasoning about sequential programs, where we know that instructions may be reordered, but the result is the same as if they executed in program order. This intuition is no longer valid for multithreaded execution. The ordering of instructions executed by multiple threads is governed by the memory model of the system.

Memory Model

A useful way to look at multithreaded execution is to imagine they are all running on a single core, so they have to be executed in sequence in some order. An example of such an interleaved order for the sum of arrays a and b could be this:

1: $r00 \leftarrow$ load $a[0]$	**5:** $r01 \leftarrow$ load $b[0]$	**9:** $r21 \leftarrow$ add $r20, r21$	**13:** $c[0] \leftarrow$ store $r01$
2: $r10 \leftarrow$ load $a[1]$	**6:** $r11 \leftarrow$ load $b[1]$	**10:** $r01 \leftarrow$ add $r00, r01$	**14:** $c[1] \leftarrow$ store $r11$
3: $r20 \leftarrow$ load $a[2]$	**7:** $r21 \leftarrow$ load $b[2]$	**11:** $r31 \leftarrow$ add $r30, r31$	**15:** $c[2] \leftarrow$ store $r21$
4: $r30 \leftarrow$ load $a[3]$	**8:** $r31 \leftarrow$ load $b[3]$	**12:** $r11 \leftarrow$ add $r10, r11$	**16:** $c[3] \leftarrow$ store $r31$

Of course, we want to run each thread on its own core to speed up execution, but logically the result (array c) after multicore execution is the same as if they did execute in interleaved fashion on one core.

We can observe that each thread executes its instructions in program order in this example. The order of the interleaving is arbitrary, although here the loads and stores are in thread order and the adds are in a permuted order. This type of multithreaded execution order, which preserves the program order of each thread, is called *sequential consistency*. It's the most natural way to think about multithreaded execution, but it's not the way most processors work. The problem comes as a result of instruction reordering by the processor at runtime. How this is done is a complex topic which we won't get into [68]. It's only important to know that it's done to improve performance and that it can also be a threat to the correctness of parallel program execution. Reordering is limited by the dependencies between the instructions of each thread. In this example the loads of each thread can obviously be done in either order, but they must complete before the add, after which the store can take place.

Returning to our attempt to synchronize threads 0 and 3, we can see that the instructions of thread 3 could be reordered, since they each access a different memory location. This could lead to the following order:

thread 3:	$flag \leftarrow$ store 1
thread 0:	L: $r00 \leftarrow$ load $flag$
thread 0:	If $flag \neq 1$ goto L
thread 0:	$r00 \leftarrow$ load $c[3]$
thread 0:	$x \leftarrow$ store $r00$
thread 3:	$c[3] \leftarrow$ store $r31$

What we have tried to do here is to intentionally use a data race with $flag$ to synchronize two threads, which is called a *synchronization race*. However our solution has been undone by the reordering of instructions. Note that this ordering violates sequential consistency, because the instructions of thread 3 don't take effect in program order.

Sequential consistency is one possible *memory model*. Other models involve some form of *relaxed consistency* where the requirement for instructions to execute in program order is relaxed, in order to obtain increased performance.

It's possible to reason about program execution for relaxed consistency models, but it's much easier to use sequential consistency. Fortunately, we can keep reasoning with the sequential consistency model, even if the underlying memory model uses relaxed consistency, as long as we eliminate data races. If we need to synchronize the execution of threads we can only use specialized language constructs, such as locks or barriers, to perform synchronization, rather than rolling our own as in this example. This sequential consistency for data race free execution applies to higher level languages such as C++ and Java as well [68].

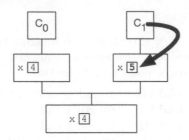

Figure 2.7: Cache coherence problem. Core C_0 loads 4 from memory location x and C_1 writes 5 to x.

Cache Coherence

The discussion so far has ignored the effect of cache memory. On-chip caches offer performance benefits by enabling much faster access to a relatively small amount of memory. The operation of most caches is invisible to programmers. A block of data is fetched from main memory into cache memory when a load or store occurs to a data element not already in the cache. Programmers concerned about performance know to maximize the number of memory accesses to cache blocks, for example by favoring traversal of consecutive elements of arrays. Otherwise, cache memory can safely be ignored for sequential programs.

In a single core processor the fact that there can be two copies of the data corresponding to the same memory address is not a problem, as the program can never access an out of date value in main memory. The situation is different in a multicore processor. As Figure 2.7 illustrates, one core has loaded the value corresponding to variable x in its own cache and the other core has updated the value of x in its cache. The first core could then read an outdated value of x. In most processors this problem is solved by using a hardware protocol so that the caches are coherent. A good definition of *cache coherence* states that at any given time either any number of cores can read or only one core can write to a cache block, and in addition that once a write occurs it is immediately visible to a subsequent read [68]. In other words cache coherence makes the cache almost disappear from the multicore programmer's view, since the behavior described in the definition is what is expected of concurrent accesses to the same location in main memory. Since loads and stores are atomic it is not possible for multiple writes (or reads and writes) to occur simultaneously to the same memory location.

Why does the cache only *almost* disappear? This is because the above definition of cache coherence refers to a *cache block*, which contains multiple data elements. If cores are accessing distinct data that are in the same cache block in their local cache they cannot do concurrent writes (or writes and reads). The programmer thinks the threads are making concurrent accesses to these data elements, but in fact they cannot if the block is in coherent caches. This is called *false sharing*, which we'll explore when we discuss barriers to performance in Chapter 5.

2.1.5 Summary

We've seen the three machine models that are relevant to parallel programmers. Each one can incorporate the previous one, as SIMD execution is enabled in most processors, and distributed multicomputers consist of a number of shared memory multiprocessors. We'll see in Chapter 4 that there are different implementation strategies for each model.

Parallelism is expressed differently in each model. SIMD execution is usually specified implicitly by the lack of dependencies in a loop. Distributed memory programming makes use of message passing, where communication is explicit and synchronization is implicitly specified by the messages. In shared memory programming, threads synchronize explicitly but communication is implicitly done by the hardware.

Program correctness is a serious issue when memory is shared. Understanding how to eliminate data races is key, since it enables us to use the intuitive sequential consistency model to reason about program execution. We'll explore synchronization techniques in Chapter 4.

Data movement is a key concern for all models. Successful implementations will minimize communication. A good way to do this is to favor access to local data over remote data.

2.2 PARALLEL EXECUTION MODEL

All programming languages have an execution model that specifies the building blocks of a program and their order of execution. Sequential execution models impose an order of execution. Compilers can reorder machine instructions only if the result is unchanged. Parallel execution models specify tasks and, directly or indirectly, the dependencies between them. The order of execution of tasks is only constrained by the dependencies. Programmers use a language's execution model as a mental model when designing, writing, and reading programs.

Execution models are also necessary for theoretical analysis of algorithms. A number of models have been proposed, which represent both a machine and an execution model. The random access machine (RAM) is a sequential model with a single processing unit and memory. Execution is measured in the number of operations and memory required. The parallel RAM (PRAM) model has multiple processing units working synchronously and sharing memory. Subcategories of the PRAM model specify how concurrent accesses to the same memory location are arbitrated.

Other theoretical models take into account the computation and communication costs of execution on parallel computers. While they can more accurately assess algorithm performance than the simple PRAM model, they are not necessarily the best suited as execution models for parallel algorithm design. It's becoming increasingly difficult to model complex parallel computers at a level relevant to programmers. A suitable model should capture the essential features of parallel algorithms at a level of abstraction high enough to be applicable for execution on any computer.

2.2.1 Task Graph Model

The *task graph* model is our candidate for an execution model. A parallel program designer is concerned with specifying the units of work and the dependencies between them. These are captured in the vertices and edges of the task graph. This model is independent of the target computer, which is appropriate for the designer since algorithm exploration should start with the properties of the problem at hand and not the computer. This model allows theoretical analysis, as we will see in Chapter 5. Of course, task graphs don't operate on their own, they just specify the work. They need to be mapped to a particular computer. The task graph model is well suited to analyzing this process as it is the preferred model for analysis of task scheduling on parallel computers.

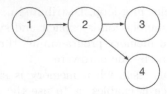

Figure 2.8: Dependence graph for Example 2.1.

Real Data Dependencies

Before we define the task graph model we need to be clear about what we mean by dependencies between tasks. Consider the following four statements:

Example 2.1

1: $a \leftarrow 5$
2: $b \leftarrow 3 + a$
3: $c \leftarrow 2 * b - 1$
4: $d \leftarrow b + 1$

Normally a programmer thinks of these statements executing as they appear in a program, that is, one after another. Let's look more closely. Statement 1 assigns a value to a, which is then needed by statement 2. Statement 2 assigns a value to b, which is needed by both statements 3 and 4. These dependencies can be represented by the graph in Figure 2.8. Now we can see that there are several possible ways these statements could be executed. Since there is no dependence between statements 3 and 4, they could be executed in either order or in parallel. The first two statements must be executed sequentially in order, however, because of the dependence between them.

The edges of the graph in Figure 2.8 also represent the flow or communication of data between statements. The value of a is communicated between statements 1 and 2, and statement 2 communicates the value of b to statements 3 and 4. In practice the communication might occur simply by writing and reading to a register, but for parallel execution it could involve sending data over a communication link.

The dependencies in Example 2.1 are called *real dependencies*. In contrast, consider the next group of statements:

Example 2.2

1: $a \leftarrow 8$
2: $b \leftarrow 7 * a$
3: $c \leftarrow 3 * b + 1$
4: $b \leftarrow 2 * a$
5: $c \leftarrow (a + b)/2$

Statements 2, 4 and 5 require the value of a to be communicated from statement 1, and statement 3 needs the value of b from statement 2. There is also a dependence between statements 3 and 4, through b, which requires them to be executed one after the other. This dependence is in the reverse direction of the other ones, since it involves a read followed by a write rather than a write followed by a read. This type of dependence is called an *anti-dependence*. It's also known as a *false dependence*, since it can easily be eliminated by using another variable in statement 4, $d \leftarrow 2 * a$, and changing statement 5 to $c \leftarrow (a + d)/2$.

There is another type of false dependence, illustrated in the following example:

Example 2.3

1: $a \leftarrow 3$
2: $b \leftarrow 5 + a$
3: $a \leftarrow 42$
4: $c \leftarrow 2 * a + 1$

Statements 1 and 2 should be able to be executed independently from statements 3 and 4, but they can't because the final value of a (42) requires statement 1 to be executed before statement 3. This is called an *output dependence*, since two statements are writing to the same variable. It is a false dependence because it can easily be eliminated by replacing a with another variable in statements 3 and 4.

The examples so far have shown data dependencies. Control dependencies are also familiar, as in:

Example 2.4

1: **if** $a > b$ **then**
2: $\quad c \leftarrow a/2$
3: **else**
4: $\quad c \leftarrow b/2$
5: **end**

Only one of statements 2 and 4 is executed. The task graph model does not model control dependencies and is only concerned with real data dependencies.

Definition 2.1 (Task Graph Model). *A parallel algorithm is represented by a directed acyclic graph, where each vertex represents execution of a task, and an edge between two vertices represents a real data dependence between the corresponding tasks. A directed edge $u \rightarrow v$ indicates that task u must execute before task v, and implies communication between the tasks. Tasks may be fine-grained, such as a single statement, or may be coarse-grained, containing multiple statements and control structures. Any control dependencies are only encapsulated in the tasks. The vertices and edges may be labeled with nonnegative weights, indicating relative computation and communication costs, respectively.*

We'll see in the following chapters that the granularity of tasks is an important consideration in parallel algorithm design. The rest of this chapter explores some simple examples of task graphs generated from sequences of statements, to get a feeling for how this execution model represents dependencies between tasks.

2.2.2 Examples

Work-intensive applications usually spend most of their time in loops. Consider the loop for addition of two arrays in Example 2.5.

Example 2.5

```
for i ← 0 to n − 1 do
    c[i] ← a[i] + b[i]
end
```

The statement in each iteration can be represented by a vertex in the task graph, but no edges are required since the statements are independent (Figure 2.9a). We are not usually this lucky, as often there are dependencies between iterations, as in Example 2.6.

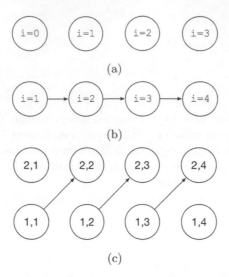

Figure 2.9: Task graphs for four iterations of Examples (a) 2.5, (b) 2.6 and (c) 2.7.

Example 2.6

> **for** $i \leftarrow 1$ **to** $n - 1$ **do**
> $a[i] \leftarrow a[i - 1] + x * i$
> **end**

Each iteration depends on the value computed in the previous iteration, as shown in Figure 2.9b for the case with four iterations.

Task graphs for iterative programs can get very large. The number of iterations may not even be known in advance. Consider Example 2.7:

Example 2.7

> **for** $i \leftarrow 1$ **to** $n - 1$ **do**
> 1: $a[i] \leftarrow b[i] + x * i$
> 2: $c[i] \leftarrow a[i - 1] * b[i]$
> **end**

Figure 2.9c shows the task graph for the first four iterations of Example 2.7. Each task has an extra label indicating to which iteration it belongs. The dependence between tasks 1 and 2 is from each iteration to the next.

There are two alternative ways to model iterative computation. One is to only show dependencies within each iteration, which would produce two unconnected tasks for Example 2.7. Another alternative is to use a compact representation called a *flow graph*, where tasks can be executed multiple times, which we won't discuss here.

Reduction

One particular case is interesting, as it represents a common pattern, called a *reduction*. We'll bring back the word count example from Chapter 1, but to simplify we are only interested in occurrences of a given word:

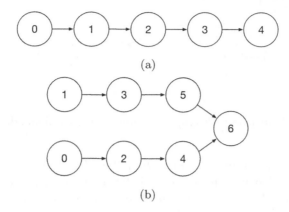

Figure 2.10: Task graph for (a) sequential and (b) parallel reduction.

count ← 0
foreach *document in collection* **do**
    count ← count + countOccurrences(word, document)
end

Let's see if we can build a task graph with independent tasks, which will allow parallel execution. To keep things simple, we'll take the case where there are four documents, and unroll the loop:

0: count ← 0
1: count ← count + countOccurrences(word, document1)
2: count ← count + countOccurrences(word, document2)
3: count ← count + countOccurrences(word, document3)
4: count ← count + countOccurrences(word, document4)

These statements aren't independent, since they are all updating the count variable, as shown by the task graph in Figure 2.10a. This contradicts our intuition, which tells us that we can compute partial sums independently followed by a combination of the result, as in:

0: count1 ← 0
1: count2 ← 0
2: count1 ← count1 + countOccurrences(word, document1)
3: count2 ← count2 + countOccurrences(word, document2)
4: count1 ← count1 + countOccurrences(word, document3)
5: count2 ← count2 + countOccurrences(word, document4)
6: count ← count1 + count2

As the task graph in Figure 2.10b illustrates, statements 0, 2 and 4 can be computed at the same time as statements 1, 3, and 5. The final statement needs to wait until tasks 4 and 5 are complete. We'll look at reduction in more detail in Chapter 3.

Transitive Dependencies

We've left out some dependencies between the tasks for sequential reduction in Figure 2.10a. Each task writes to *count* and all following tasks read from *count*. In Figure 2.11 we show

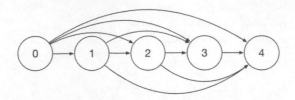

Figure 2.11: Reduction task graph showing transitive dependencies.

all the real dependencies. This graph illustrates *transitive dependence*. For example, task 1 depends on task 0, task 2 depends on task 1, and the transitive dependence where task 2 depends on task 0 is also included. The process of removing the transitive dependencies to produce Figure 2.10a is called *transitive reduction*.

We don't always want to remove transitive dependencies, as in the following example:

Example 2.8

1: $y \leftarrow$ foo()
2: $ymin \leftarrow$ min(y)
3: $ymax \leftarrow$ max(y)
4: for $i \leftarrow 0$ *to* $n - 1$ do
    $ynorm[i] \leftarrow (y[i] - ymin)/(ymax - ymin)$
end

Statement 1 generates an array y, statements 2 and 3 find the minimum and maximum of y and the loop uses the results of the three previous statements to produce array *ynorm* with values normalized between 0 and 1. The tasks graph is shown in Figure 2.12, with the edge labels indicating the relative size of the data being communicated between tasks. Here the transitive dependence between tasks 1 and 4 is useful as it not only indicates the dependence of task 4 on task 1 (through y) but also that this dependence involves a communication volume that is n times larger than its other dependencies.

Mapping Task Graphs

Task graphs are a very useful tool for parallel algorithm design. An important design goal is to have many tasks with as few dependencies as possible. We'll see in Chapter 5 that a task graph can be analyzed to assess properties such as the available parallelism.

The task graph then has to be mapped to a given machine, whether this is done by the programmer or by the runtime system. Edges between tasks on different processors indicate

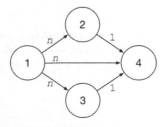

Figure 2.12: Task graph for Example 2.8.

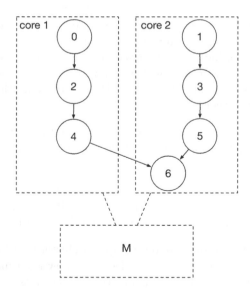

Figure 2.13: Reduction task graph mapped to dual core processor.

the need for communication and/or synchronization, and add overhead to the execution time of the program. Take for example the mapping in Figure 2.13 of the reduction task graph in Figure 2.10b to a machine with two cores sharing memory. The dependency between statements 4 and 6 means that before executing statement 6, core 2 will need to wait until core 1 has finished with statement 4, and then obtain the value of variable *count1*.

2.2.3 Summary

If all that was required to design a parallel algorithm was to do a dependency analysis on the steps of a sequential algorithm there would be no need for parallel programming. As this reduction example illustrates, at a minimum tasks need to be rewritten to expose sufficient parallelism. Parallel algorithms can differ significantly from their sequential counterparts.

We'll explore algorithmic and implementation building blocks in the next two chapters. The goal is to develop intuition about parallel computing and to provide a collection of structures that can be assembled into complex parallel algorithms.

2.3 FURTHER READING

Any good textbook on computer architecture, such as Patterson and Hennessy's *Computer Organization and Design* [54], will include an overview of multiprocessors and multicomputers. A primer on SIMD execution can be found in Hughes's *Single-Instruction Multiple-Data Execution* [40]. The growing importance and computing power of multiprocessors is reflected in the literature, such as Solihin's *Fundamentals of Parallel Multicore Architecture* [67]. *A Primer on Memory Consistency and Cache Coherence* [68] is worth exploring for a deeper understanding of memory models. The task graph is used as an execution model for multi-threaded execution in the third edition of Cormen et al.'s *Introduction to Algorithms* [13]. An excellent presentation of graph models of execution and their use in scheduling analysis is given in Sinnen's *Task Scheduling for Parallel Systems* [65].

2.4 EXERCISES

2.1 Look up the technical details of two processors that include SIMD execution units. Report the basics of the microarchitecture, including vector registers, ALUs, and vector instructions. Assuming ideal independent execution on contiguous data, such as the array addition example in section 2.1.1, how much speedup can SIMD execution provide on each processor? Contrast the two SIMD implementations. Will one always perform better than the other?

2.2 Look up the details of the memory hierarchy of two multicore processors. Sketch the memory layout of each processor and report the capacity of each level of cache/memory.

2.3 Describe the architecture of a NUMA multiprocessor. Include details of the network that connects the memory modules of each processor. If you can, find a typical access time and/or bandwidth difference for local and non-local loads/stores.

2.4 Look up details of an institutional shared computer cluster that incorporates all three machine models described in this chapter. Describe how each machine model is implemented. Give a description of the network that connects the compute nodes.

2.5 For the cluster found for exercise 2.4 describe how users obtain an account and how they can run their programs on the cluster. Discuss how using such a shared cluster differs from using a personal computer.

2.6 Trace the following pseudocode:

$$b \leftarrow [4, 3, 8, 9]$$
$$c \leftarrow [3, 1, 3, 2]$$
for $i \leftarrow 0$ *to* 3 **do**
$\quad a[c[i]] \leftarrow b[i]$
end

What would be the result if the loop was executed sequentially? What would be the result if each iteration of the **for** loop were executed in parallel by a different thread?

2.7 Construct a task graph to describe cleaning a classroom, including sweeping and washing the floors, cleaning desktops, etc. Create tasks in order to enable parallel execution. Pay careful attention to the dependencies between tasks. Given your task graph, how many people could productively help accomplish the task faster?

2.8 a. Draw a task graph for the inner loop of the bubble sort algorithm. Are there independent tasks that could be executed in parallel?

```
for i ← 0 to n − 2 do
    for j ← n − 1 down to i + 1 do
        // swap elements if first greater than second
        CompareSwap(a[j − 1], a[j])
    end
end
```

b. The *odd-even transposition sort* is a variation on bubble sort. Draw the task graph for the inner loops. Are there independent tasks that could be executed in parallel?

```
for i ← 0 to n − 1 do
    if i is even then
        for j ← 0 to ⌊n/2⌋ − 1 do
            CompareSwap(a[2j], a[2j + 1])
        end
    end
    if i is odd then
        for j ← 1 to ⌈n/2⌉ − 1 do
            CompareSwap(a[2j − 1], a[2j])
        end
    end
end
```

CHAPTER 3

Parallel Algorithmic Structures

This chapter examines common patterns in parallel algorithm design, in the form of task graph structures. The fundamental activity of parallel algorithm design is *decomposition*, which means identifying tasks and their dependencies. We begin with a simple example.

3.1 HISTOGRAM EXAMPLE

We have a very large collection of short messages from a social network and we want to produce a histogram showing the distribution of the complexity of the messages. Assume that the complexity can be estimated as the fraction of the letters in the alphabet that occur in a message, and that our histogram will contain 16 bins covering the range of complexities from 0 to 1. The first step is to calculate the complexity of each message. In the second step the complexity value of each message needs to be placed in one of the histogram bins. For the first step we can assign the calculation of the complexity of one message to a task, and as a result have as many tasks as messages, with tasks working independently.

Let's consider two candidate decompositions for the second step. We can choose to have a task examine all complexity values from the first step and fill the contents of one bin. This results in 16 tasks working independently, but with a dependence on the results of the first set of tasks. We then need to add a third step with one task to gather the results from these 16 tasks to produce a histogram. The resulting task graph is shown in Figure 3.1a. This decomposition is obviously limited to only 16 parallel tasks in the second step.

An alternative decomposition is to have each task create a histogram for a message. This means that each histogram will only have a single value in one bin. This requires a third step as in the previous decomposition, this time to merge the histograms from the second step, as seen in Figure 3.1b. As we'll see below, this merge step is a common pattern called a reduction. It may seem strange to create so many lightweight tasks, but it is very important to begin by discovering as much parallelism as possible. Tasks can be agglomerated to improve performance, given the characteristics of the target machine, as we'll see in the next chapter.

An important property of a decomposition is how the number of tasks increases with the problem size. In the first level of both decompositions the number of tasks increases with the number of messages. This is obviously desirable, because it means we can marshal more computing resources as the problem size increases. The situation is different at the second

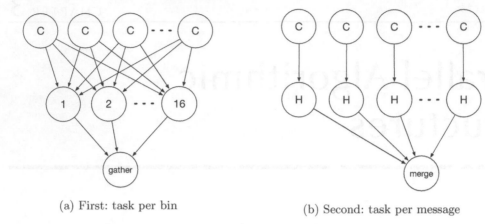

      (a) First: task per bin   (b) Second: task per message

Figure 3.1: Two decompositions for histogram problem.

level. The workload of tasks in the second level in Figure 3.1a increases with the number of messages and the number of tasks stays constant. In contrast, the workload of the second level tasks in Figure 3.1b stays the same, but the number of tasks increases as the number of messages increases.

3.1.1 Guidelines for Parallel Algorithm Design

It's important to emphasize that in designing the task graphs above we have not said anything about the number of cores that will be used. This is good practice, as stated in the first guideline.

Guideline 3.1 *Postpone consideration of the details of the computational platform until after the decomposition phase.*

When performing a decomposition, resist the temptation to think of tasks as cores. The second guideline addresses the scalability of the decomposition.

Guideline 3.2 *Create many independent tasks, whose number increases with the problem size.*

These guidelines can be found in Foster's design methodology for parallel programming from the mid '90s [24]. They are becoming increasingly relevant, as parallel computers are becoming more complex and are evolving rapidly. A design that is tied to a particular architecture is unlikely to perform well on other machines, whereas an abstract design based on tasks and dependencies can be adapted to run on any machine. An efficient mapping to a computer with a multilevel hierarchy of heterogeneous functional units can be very hard to do. This is why semi-implicit programming models (see Chapter 1) are appealing, as they take care of the mapping of an abstract decomposition.

Applying these principles to our simple example reveals the superiority of the second decomposition. There are more tasks, and their number increases with the problem size. This means that there is more potential parallelism. Of course we will want to have fewer tasks, with each handling a subset of messages, but our goal here is to identify as much parallelism as possible.

The following structural patterns arise frequently in parallel algorithm design. Studying

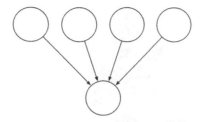

Figure 3.2: Task graph for embarrassingly parallel pattern.

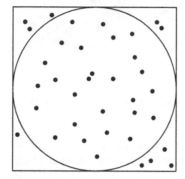

Figure 3.3: Monte Carlo estimation of π.

them can deepen our understanding of decomposition. They are often used as components of parallel algorithms.

3.2 EMBARRASSINGLY PARALLEL

This pattern occurs quite frequently, and is nothing to be embarrassed about. A better term might be *obviously parallel*, as it applies to cases where the decomposition into independent tasks is self-evident. Each task does its own computation independent of the other tasks, and the resulting task graph is completely disconnected. The definition of embarrassingly parallel excludes input of initial data and collection of results at the end of execution. The task graph for this pattern typically looks like Figure 3.2. This graph includes the collection of results, but the bulk of the computational work takes place in the independent tasks on the first level of the graph.

In Monte Carlo methods, for example, a problem is solved by conducting numerous experiments with pseudo-randomly selected variables. These methods can be used to numerically evaluate integrals that arise in areas such as physics and finance. The classic illustrative example is the approximation of π. A circle of radius 1 is inscribed in a square and points are randomly chosen inside the square (see Figure 3.3). The area of the circle, and hence π, can be estimated as the fraction of points that lie in the circle multiplied by the area of the square. This problem can be decomposed into independent tasks that each generate a point and determine whether it lies in the circle.

Another frequently cited example is the generation of fractal images. We will look at a generalization of the Mandelbrot set, which produces even more interesting images. Generalized fractals are generated through the function $z \leftarrow z^\alpha + c$, where α is a real number and z is a complex number with initial value $(0,0)$ for $\alpha > 0$ and $(1,1)$ for $\alpha < 0$ [32].

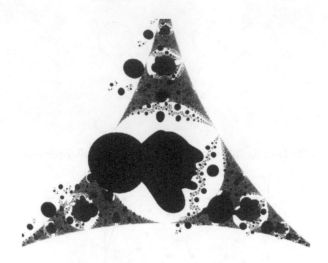

Figure 3.4: Generalized fractal for $\alpha = -2$.

The parameter α is chosen for a given image, and the function is iterated for values of c whose real and imaginary parts refer to coordinates of a pixel in the image. The iteration frequently diverges, and the number of iterations required to diverge is mapped to a color. For a grayscale image, white pixels represent iterations that don't diverge after 255 iterations, and for others the darker the gray the earlier the iteration diverges. A fractal for $\alpha = -2$ is shown in Figure 3.4. One can assign a task to computation of a pixel, and clearly all tasks are independent. Notice, though, that the computational cost of the tasks vary depending on the number of iterations taken. This will be important when assigning tasks to processing units, as the workload imbalance is an issue. Tasks computing black pixels wait while those computing white pixels complete the maximum number of iterations. This can be addressed by using a dynamic load balancing technique, as we'll see in Chapter 4.

3.3 REDUCTION

Reduction is the most commonly encountered pattern. The success of frameworks based on Google's MapReduce is a result of the ubiquity of the reduction pattern. It arises when an associative binary operation is performed on a collection of data. It includes arithmetic and logical operations such as sum and maximum. Reduction is accomplished sequentially using a `for` loop. As we saw in Chapter 2, if the results are accumulated in a single variable the iterations cannot be evaluated independently. However, the associativity of the operation allows us to do partial accumulations independently in any order.

We can assign a task to the evaluation of one operation. We first add disjoint pairs of values. This means that for n values we can do $n/2$ tasks in parallel, resulting in a halving of the number of values to process. We can then assign another $n/4$ tasks, where each task again computes one operation. This continues in stages until the number of values is reduced to one, forming a binary tree, as illustrated in Figure 3.5a. Note that this is a complete binary tree since the number of values to be reduced is a power of two. This decomposition works equally well if the number of of values is not a power of two, and the resulting binary tree is not complete (see Figure 3.5b for 7 values).

The binary tree decomposition creates $n/2^i$ tasks that can be executed in parallel at the ith step. It has the virtue that the number of tasks is proportional to the size of the data.

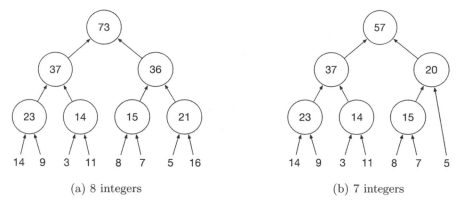

(a) 8 integers   (b) 7 integers

Figure 3.5: Sum reduction.

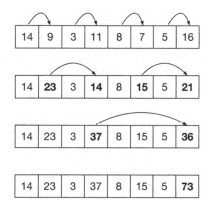

Figure 3.6: Data structure view of reduction from Figure 3.5a.

However, it suffers from the halving of the number of tasks at each step. This means that the number of tasks executing in parallel decreases as the computation proceeds.

The task graph representations in Figure 3.5 don't show anything about how the values are laid out in memory. For a distributed memory implementation the values could be scattered among processes. For a shared memory or SIMD implementation the values would likely be stored in an array. Figure 3.6 is a representation of the tasks in Figure 3.5a acting on an array, where the array is overwritten and the result stored in the final element. The bold values indicate the result of the tasks in the previous level. We are free to reorder the leaves in Figure 3.5a, which would modify the memory access pattern. The value of the representation in Figure 3.6 is that it shows the pattern of loads and stores in memory. Chapter 4 will show how the order of memory accesses affects the performance of SIMD implementation of reduction.

The values to be reduced can also be data structures such as arrays, as the next example illustrates.

K-means Clustering

A practical application of this pattern can be found in the k-means clustering algorithm. This algorithm is used to assign points in n-dimensional space to k clusters. It arises in

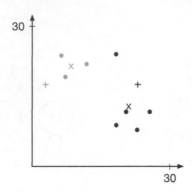

Figure 3.7: K-means example with 8 points and two clusters. Centers marked with + for initial guess and × for centroid of points assigned to cluster. In the next iteration the point in the upper right quadrant will be assigned to the gray cluster.

applications such as data mining, where one has a large collection of observations with n attributes, and one wants to group them into a number of clusters. A sketch of the algorithm is given in Algorithm 3.1.

Algorithm 3.1: K-means algorithm

Input: array of n-dimensional vectors, and the number of clusters (k)
Output: array *closest* containing assignment of each vector to one of the clusters

1: Assign initial guesses of the k cluster centers $cluster[k]$
2: Initialize $clusterNew[k]$ and $clusterSize[k]$ arrays to zero
3: **while** *not converged* **do**
4:　　**foreach** *vector j* **do**
5:　　　　find cluster center i with smallest Euclidean distance to vector j
6:　　　　$closest[j] \leftarrow i$
7:　　　　$clusterNew[i] \leftarrow clusterNew[i] + vector[j]$
8:　　　　$clusterSize[i] \leftarrow clusterSize[i] + 1$
9:　　**end**
10:　　**foreach** *cluster center i* **do**
11:　　　　$cluster[i] \leftarrow clusterNew[i]/clusterSize[i]$
12:　　　　$clusterNew[i] \leftarrow 0$
13:　　　　$clusterSize[i] \leftarrow 0$
14:　　**end**
15: **end**

The outer iteration converges when the number of changes in cluster membership falls below a set threshold. We can assign an independent task to each iteration of the `foreach` loop in lines 4–9. Each task takes one vector, finds the closest (in terms of Euclidean distance) cluster center, stores its vector in the corresponding element of a k-dimensional array *clusterNew*, and increments the count of vectors assigned to that cluster.

Consider the very simple example with two clusters in Figure 3.7. Each of 8 tasks would produce a *clusterNew* array:

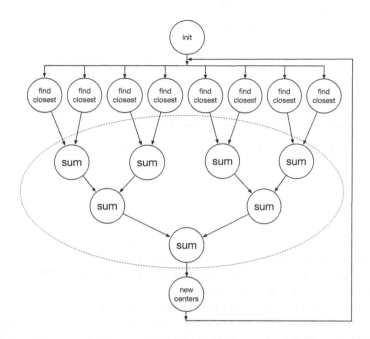

Figure 3.8: Task graph for k-means algorithm. Reduction indicated by dashed oval.

$$[[6.2, 23.9], 0] \qquad [[7.1, 19.3], 0] \qquad [[11.7, 22.1], 0] \qquad [0, [18.1, 24.2]]$$
$$[0, [18.1, 8.9]] \qquad [0, [20.3, 11.8]] \qquad [0, [22.7, 7.9]] \qquad [0, [25.4, 11.8]]$$

a *clusterSize* array:

$$[1, 0] \qquad\qquad [1, 0] \qquad\qquad [1, 0] \qquad\qquad [0, 1]$$
$$[0, 1] \qquad\qquad [0, 1] \qquad\qquad [0, 1] \qquad\qquad [0, 1]$$

and update one element of the array $closest = [0, 0, 0, 1, 1, 1, 1, 1]$.

Next the coordinates of vectors and counts need to be summed for each cluster. This is accomplished by performing a sum reduction of the *clusterNew* and *clusterSize* arrays, which results in *clusterNew* $= [[25.0, 65.3], [104.6, 64.6]]$ and *clusterSize* $= [3, 5]$. Finally, a single task calculates the new centroid of each cluster (line 11), *cluster* $= [[8.3, 21.8], [20.9, 12.9]]$. This example converges in the following iteration where each cluster is assigned 4 members. The task graph is shown in Figure 3.8.

3.4 SCAN

The scan operation is closely related to a reduction. It applies a binary associative operator to an *ordered* collection of values, resulting in a new collection where element i contains the result of the operator applied to the first i elements of the original collection. When the operation is a sum, a scan is called a *prefix sum*, since the sum is taken of all prefixes of the collection. For instance, if we have the array $[2, 16, 9, 7]$, the prefix sum is $[2, 18, 27, 34]$. This type of scan, which includes the ith value in the sum for the ith element, is called an *inclusive scan*. One can also define an *exclusive scan*, where the ith value is excluded; for our example this gives $[0, 2, 18, 27]$.

Scan has quite a long history. It was first proposed as an operator for the APL programming language in the 1950s [7]. It is also a very useful primitive for many parallel

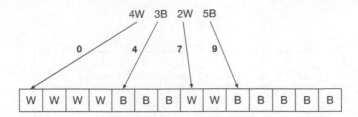

Figure 3.9: Parallel run-length decoding using exclusive prefix sum.

algorithms, including radix sort, quick sort, and sparse matrix-vector multiplication. We'll see it again in later chapters on the Single Source Shortest Path problem (Chapter 6) and the computation of the planar convex hull (Chapter 8).

A simple example involves run-length decoding. Run length encoding compresses a stream of data that includes many repeated adjacent values. A black and white image might include pixels in the following sequence: WWWWBBBWWBBBBB. Each repeating sequence can be replaced by a pair representing the frequency and color, which would be 4W3B2W5B. The decoding operation simply replaces each pair with the specified number of repeated characters. One can clearly expand each pair independently, but if the pairs are expanded in parallel how do we know where in the array to place each expanded sequence of characters? A prefix sum of the frequencies gives us this information. In our example, the exclusive prefix sum gives $[0, 4, 7, 9]$, which indicates the starting index of each expansion, as seen in Figure 3.9.

Parallel scan is not as straightforward as a parallel reduction. One can see this from the sequential pseudocode:

$sum \leftarrow 0$
for $i \leftarrow 0$ *to* $n - 1$ **do**
    $sum \leftarrow sum + a[i]$
    $scan[i] \leftarrow sum$
end

For the parallel reduction we could easily see how to compute sums independently. Here the value of $scan[i]$ depends on the value of $scan[i-1]$, so it looks like the elements of $scan$ can't be computed independently. But it is possible, with some ingenuity. We'll look at two ways we can do a scan using independent tasks. The first is an extension by Blelloch of the binary tree reduction [7].

Look again at the reduction tree in Figure 3.5a. It was used to compute the overall sum of 73, but it also contains values needed to construct the prefix sum. These values can be used in a pass back down the tree. We first clear the value the root contains to zero. The root then passes its value to its left child, and to its right child it passes its value plus the partial sum its left child had in the initial pass up the tree, and so on (see Figure 3.10):

$scan[root] \leftarrow 0$
$scan[left[v]] \leftarrow scan[v]$
$scan[right[v]] \leftarrow scan[v] + reduce[left[v]]$

where $reduce$ is the partial sum produced in the first pass. As in the reduction above, the number of values does not have to be a power of two. We'll see in the next chapter that one can adapt this algorithm so that there are far fewer than $n - 1$ tasks.

The second parallel scan algorithm, by Hillis and Steele [37], is more easily understood by looking at operations on an array. Figure 3.11 shows that at iteration i the value at index

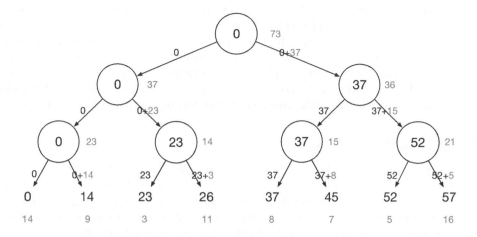

Figure 3.10: Parallel scan with two passes of binary tree, showing the second pass down the tree. The first pass up is the reduction shown in Figure 3.5a. The results of the reduction are shown as gray labels to the right of the vertices. The edge labels indicate the value passed to a child vertex. The original values are shown in the bottom row in gray.

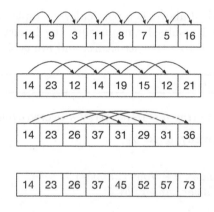

Figure 3.11: Hillis and Steele parallel scan of an array.

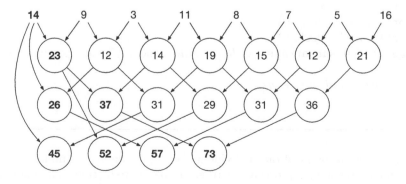

Figure 3.12: Task graph of Hillis and Steele parallel scan.

$i < n - 2^i$ is added to the value at index $i + 2^i$. Note that this performs an *inclusive scan*, compared to the exclusive scan produced by the previous algorithm. It is trivial to convert between inclusive and exclusive scans. The task graph representation of this algorithm is shown in Figure 3.12. The resulting values, shown in bold, are held by tasks that depend directly on the first value (14) and descendants of these tasks. We can readily observe that the number of levels is the same as Blelloch's algorithm, but there are more parallel tasks at each level. The performance of these two algorithms will be compared in Chapter 5.

3.5 DIVIDE-AND-CONQUER

Divide-and-conquer is a commonly used algorithmic design strategy. It recursively breaks down a problem into smaller instances of the same problem, solves the base cases and combines them to form a solution to the whole problem. There have been many applications of this strategy, including sorting algorithms, matrix multiplication, polynomial multiplication, and computation of the convex hull. There are three types of work required:

1. splitting into subproblems

2. solving bases cases

3. combining solutions

Binary tree reduction, as described above, is a particularly simple example. The only work required is in combining partial results by evaluating the binary operator. Divide-and-conquer algorithms naturally lead to a tree structure, and are typically solved recursively.

Algorithm 3.2: Sequential merge sort

Input: array a of length n.
Output: array b, containing array a sorted. Array a overwritten.

arrayCopy(a, b) // copy array a to b
mergeSort(a, 0, n, b)

// sort elements with index $i \in [lower..upper)$
Procedure mergeSort(a, *lower*, *upper*, b)
    if $(upper - lower) < 2$ **then**
        return
    end
    $mid \leftarrow \lfloor (upper + lower)/2 \rfloor$
    mergeSort(b, *lower*, *mid*, a)
    mergeSort(b, *mid*, *upper*, a)
    // merge sorted sub-arrays $i \in [lower..mid)$ and $i \in [mid..upper)$ of a
        into b
    merge(a, *lower*, *mid*, *upper*, b)
    return
end

At first glance, the decomposition of divide-and-conquer appears simple. This would be true if all divide-and-conquer problems were as simple as reduction, but unfortunately they're not. Let's take merge sort as a representative example. Algorithm 3.2 performs a recursive sequential merge sort, alternating between two copies of the array. It overwrites a

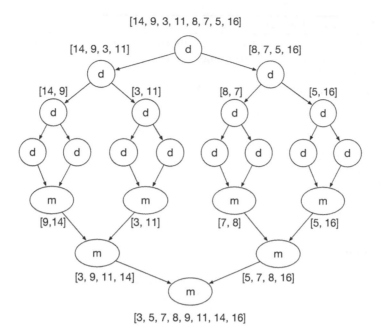

Figure 3.13: Divide-and-conquer merge sort, with divide (d) and merge (m) tasks. Each of the merge tasks can be expanded into a task graph, as in Figure 3.16.

and puts the result in b. This algorithm avoids unnecessary copying and limits the amount of extra memory used.

The obvious way to decompose this algorithm is to assign a task to each recursive instance of mergeSort. This produces a binary tree, as with reduction. An example is shown in Figure 3.13. Stopping at this point will result in poor performance, however. In the case of reduction the only work involved is in combining solutions (a single binary operation), and this work is the same for every task in the tree. The work involved in merge sort is also concentrated in the the combining of solutions, but here the work is both more complex than reduction, $O(n)$ rather than $O(1)$, and varies with the depth of the tree. Recall Guideline 3.1 that recommends maximizing the number of independent tasks. We can push the decomposition further by examining the merge operation. Algorithm 3.3 shows the sequential merge of two sorted subarrays $a[0..mid-1]$ and $a[mid..upper-1]$ into array b.

It's not easy to see how to decompose this procedure into independent tasks. The problem is that if we create a task for each placement of an element of a in b, how do we know where in b it should go? We can solve this problem by applying a divide-and-conquer strategy [13]. This strategy isn't suggested by the sequential algorithm, which naturally uses a loop, but by the goal of decomposing the merge operation into independent tasks.

Consider performing two merges independently, by dividing each subarray of a in two, as shown in Figure 3.14. Note that each half of the array is sorted in ascending order. The first task merges the first part of each subarray and the second task merges the second part. The first task places elements at the beginning of array b and the second task starts placing them at the index given by the sum of the length of the two parts the first task merges. We can't simply pick the middle of both arrays as dividing points. That would result in one task merging $[3, 9]$ and $[5, 7]$ and the other task merging $[11, 14]$ and $[8, 16]$, to produce

---

Algorithm 3.3: Sequential merge

Input: array $a[lower..upper - 1]$, with each subarray $i \in [lower..mid)$ and
$\qquad i \in [mid..upper)$ sorted in ascending order.
Output: sorted array b
Procedure merge(a, $lower$, mid, $upper$, b)
$\qquad i \leftarrow lower$
$\qquad j \leftarrow mid$
$\qquad k \leftarrow 0$
$\qquad$**for** $k \leftarrow lower$ *to* $upper - 1$ **do**
$\qquad\qquad$**if** $i < mid \wedge (j \geq upper \vee a[i] \leq a[j])$ **then**
$\qquad\qquad\qquad b[k] \leftarrow a[i]$
$\qquad\qquad\qquad i \leftarrow i + 1$
$\qquad\qquad\qquad k \leftarrow k + 1$
$\qquad\qquad$**else**
$\qquad\qquad\qquad b[k] \leftarrow a[j]$
$\qquad\qquad\qquad j \leftarrow j + 1$
$\qquad\qquad\qquad k \leftarrow k + 1$
$\qquad\qquad$**end**
$\qquad$**end**
end

$[3, 5, 7, 9, 8, 11, 14, 16]$. The elements the first task merges must be less than or equal to those of the second task, otherwise the resulting array won't be sorted.

Instead, we create a new task to split one subarray of a, say the lower one, based on its mid point, $a[mid1]$, where $mid1 = \lfloor (mid - 1 + lower)/2 \rfloor$. We need to find the portion of the upper subarray of a that is less than or equal to $a[mid1]$, which can be done using a binary search, as illustrated in Figure 3.15. This gives us the index $mid2$, which refers to the first index where the element is larger than $a[mid1]$. Then the first merge task merges elements $a[0..mid1]$ and $a[mid..mid2 - 1]$ and the second task merges elements $a[mid1 + 1..mid - 1]$ and $a[mid2..upper - 1]$. For the example in Figure 3.14 $mid1 = 1$, $mid2 = 7$, and the second merge starts placing elements at index $i = 2 + 3 = 5$. In general the "lower" half is always chosen to be the longer half, which takes care of the situation when one of the subarrays has no elements.

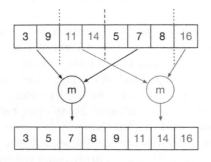

Figure 3.14: Merging with two tasks.

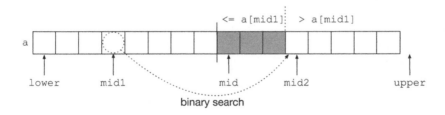

Figure 3.15: Splitting of each sorted subarray.

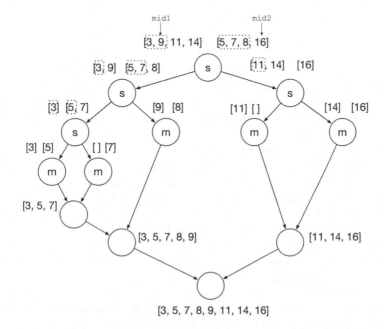

Figure 3.16: Divide-and-conquer merge, including split (s) and merge (m) tasks.

This gives us three tasks: one to split the subarrays and two to merge. We can create even more merge tasks by identifying more subarrays that can be merged in parallel. This is done by recursively decomposing the splitter task until one of the array pieces has one element or less. Then the merges take place in parallel by going back up the recursion tree. An example task graph is shown in Figure 3.16. The tasks in the final three levels represent only the return from recursive calls in a shared memory implementation, or the gathering of results in a distributed memory implementation.

The unfolding of the parallel merge sort is predictably a balanced tree (or almost balanced for n not a power of 2), whereas the merge tree unfolds in an irregular data-dependent shape. We'll see a fork-join implementation of this merge sort algorithm in Chapter 4.

3.6   PIPELINE

Consider the following loop:

Example 3.1

    while token *in input* **do**

1:      $x \leftarrow$ phase1(token)

2:      $y \leftarrow$ phase2(x)

3:      $z \leftarrow$ phase3(y)

      output z

    end

and the corresponding task graph for four iterations in Figure 3.17. We can look at executing the tasks in this graph in two ways. First, each row of tasks, corresponding to an iteration of the loop, can be executed in parallel. We can also notice that all the tasks in each column do the same computation, but on different data. Therefore the tasks in each column can all be assigned to one core. The cores form a pipeline, where each core is specialized for one type of computation. Each core iterates through its tasks, waiting for the previous core in the pipeline until its input is available. In the first step core 0 executes task $(1, 0)$. In the next step core 1 takes the result of task $(1, 0)$ and executes tasks $(2, 0)$, while at the same time core 0 executes task $(1, 1)$. In the third step all three cores are active executing tasks $(1, 2)$, $(2, 1)$, and $(3, 0)$ respectively.

The analogy usually given is a factory assembly line, but it is familiar to programmers in execution of command line statements in Unix, where commands are chained together with the pipe (|) operator, and each command operates on a stream of data, reading from standard input and writing to standard output. Pipelining can benefit any application that involves computation that takes place in stages on a stream of data, such as is found in signal processing, text processing (see Figure 1.4) and graphics.

Some image processing applications modify images by applying filters. One way to do this is to generate the frequency space representation of an image by applying a 2D Fourier transform, removing some frequency components, then applying the inverse 2D Fourier transform. An example is applying a high-pass filter to sharpen an image. This can be done in a three-stage pipeline if the filter is to be applied to a collection of images. Note that all three tasks can only execute in parallel after the first two stages have finished operating on the first image and before the first stage finishes working on the last image.

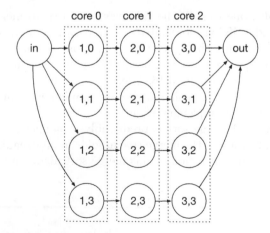

Figure 3.17: Task graph for four iterations of the loop in Example 3.1. Assignment of tasks to three cores of a pipeline is shown.

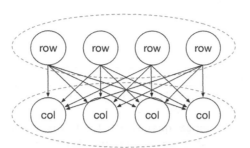

Figure 3.18: 2D FFT in two stages.

An important consideration for pipelines is the relative complexity of the tasks. Ideally all stages should take the same time to complete execution, to provide maximum overlap in time between execution of the stages. Stages that take more time become bottlenecks, as input data accumulates, and the following stage is idle as it waits for data. One way to address this issue is to enable parallel execution of time-consuming stages.

Decomposition of the stages is worth considering in all cases, since it increases the number of tasks. Consider the 2D Fourier transform of an image. It is computed by applying the 1D Fast Fourier transform to each column, then to each row. One can view this as a two-stage pipeline. We can further decompose each task by observing that the 1D Fourier transforms of each row or column can be computed independently, producing the task graph in Figure 3.18. When there is a sequence of images to be filtered there are two levels of concurrency: a coarse-grained two-stage pipeline, and a fine-grained decomposition into 1D FFTs. Considered separately, the row or column FFTs are an example of embarrassing parallelism, but the two pipeline stages are tightly coupled, as all the tasks of the second stage depend on all results of the first stage. We can actually go further with the decomposition, as parallel execution of the 1D Fast Fourier transform is well known.

Pipelined Merge Sort

The previous example illustrates how pipeline execution can be exploited as part of a parallel algorithm. A pipeline can also be used to perform merge sort in parallel. Look again at the merge sort tree in Figure 3.13. The merges take place in the final 3 (or $\log n$) levels. All merge tasks take two inputs of length 2^i and produce one output of length 2^{i+1}, $i \in [0.. \log n]$. The pipelined merge sort agglomerates tasks in each level into one task, and arranges these new tasks in a pipeline [1]. Unlike in the divide-and-conquer algorithm, where parallel merges occur at each level, here the merges are sequential, but there are $\log n$ merges taking place simultaneously when the pipeline is full. Tasks work synchronously, processing one value at each step.

The pipelined merge is detailed in Algorithm 3.4. Tasks start each iteration at the same time. Tasks have two queues where they place their input, referred to by $dir = -1$ or 1. Tasks start processing their queues when one queue is full (has 2^i elements) and the other has a single element. They then merge elements from both queues, in groups of 2^{i+1} elements. The progress through each group is tracked using nup and $ndown$ to count the number of elements selected from each queue. Input values from the previous task are placed alternatively in each queue in groups of 2^i elements at the end of each iteration.

An example is shown in Figure 3.19–3.20. In the first two iterations task 0 places the first two values in its upper and lower queues. After this point task 0 begins merging. It

---

Algorithm 3.4: Pipelined merge sort

// $m = \log n$ synchronous tasks with index $i \in [0..m)$
1: **Procedure** Task i ()
2: $dir \leftarrow 1$ // 1 points to upper queue, -1 to lower queue
3: $start \leftarrow 0$ // task can't start merging until enough values in queues
4: $count \leftarrow 0$
5: $nup \leftarrow 0$
6: $ndown \leftarrow 0$
7: **while** value *in input or queues not empty* **do**
8:      **if** $(start = 0) \wedge (upper\ queue\ has\ 2^i\ elements\ and\ lower\ has\ 1\ element)$ **then**
9:          $start \leftarrow 1$
10:     **end**
11:     **if** $start = 1$ **then**
12:         **if** $nup = 2^i \wedge ndown = 2^i$ **then**
13:             $nup \leftarrow 0$
14:             $ndown \leftarrow 0$
15:         **end**
16:         **if** $nup = 2^i$ **then**
17:             send head of lower queue to output
18:             $ndown \leftarrow ndown + 1$
19:         **else if** $ndown = 2^i$ **then**
20:             send head of upper queue to output
21:             $nup \leftarrow nup + 1$
22:         **else if** *head of upper queue* $\geq$ *than head of lower queue* **then**
23:             send head of upper queue to output
24:             $nup \leftarrow nup + 1$
25:         **else**
26:             send head of lower queue to output
27:             $ndown \leftarrow ndown + 1$
28:         **end**
29:     **end**
30:     **if** value *exists* **then**
31:         put value in queue dir
32:         $count \leftarrow count + 1$
33:         **if** $count = 2^i$ **then**
34:             $dir \leftarrow -dir$
35:             $count \leftarrow 0$
36:         **end**
37:     **end**
38: **end**
39: **end**

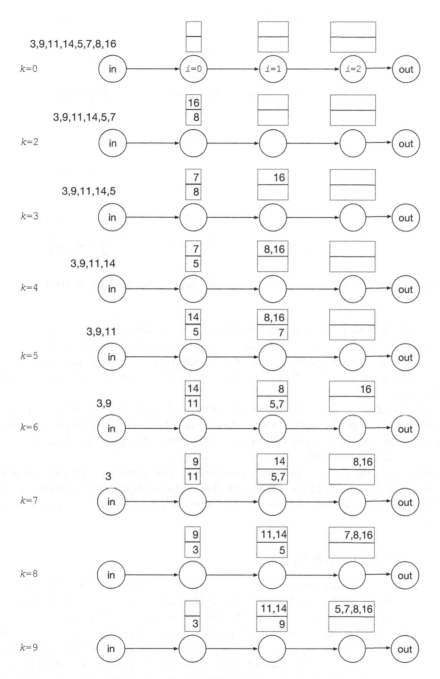

Figure 3.19: Pipelined merge sort, showing state after iterations 0 and 2–9. Boxes indicate two queues for each inner task (front of queue at the end).

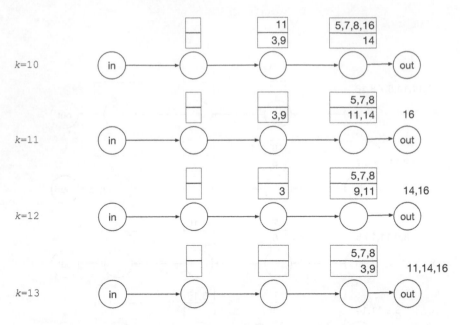

Figure 3.20: Pipelined merge sort, showing state after iterations 10–13.

sends 16 followed by 8 to the next task (iterations 3,4), 7 followed by 5 (iterations 5,6), etc. Note that tasks don't simply keep picking the maximum from each head (5 was picked in iteration 6, not 14), but merge in groups of 2^{i+1}. Once the following tasks have enough elements they too begin merging. Tasks terminate when their queues are empty and they have no more input.

Comparing this example with the merge sort graph in Figure 3.13 reveals that both parallel decompositions merge the same groups of elements. However, the divide-and-conquer algorithm merges groups of the same size in parallel, whereas the pipeline algorithm merges groups of different size in parallel, one element at a time.

This application of the pipeline pattern to merge sort illustrates the creativity of parallel algorithm design. It goes beyond the obvious divide-and-conquer parallelization. Having a diversity of algorithms is useful, as some will be more appropriate for particular machine architectures.

3.7 DATA DECOMPOSITION

So far our approach to decomposition has been focussed on tasks. We identified tasks that can be executed independently, and built task graphs taking the data dependencies into account. Another approach is to decompose the data structures, then associate a task with each portion. This is called *data decomposition*. Each task does the same work, but on distinct data. Data decomposition is most suitable when the same operation is performed on elements of a data structure. Often both task and data decomposition are used in parallel algorithms. We'll look at four different applications of data decomposition in this chapter: a grid application, matrix-vector multiplication, bottom-up dynamic programming, and pointer jumping. The next chapter will discuss a data decomposition of merge sort.

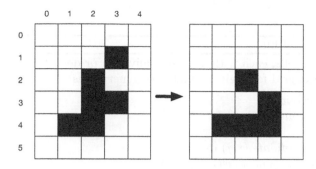

Figure 3.21: Game of Life example.

Grid Applications

A large number of applications make use of one to three dimensional grids of data. Grids arise in applications such as the numerical solution of differential equations, in graphics applications that operate on grids of pixels, and in cellular automata simulations. The computation sweeps across the grid, either overwriting elements during the sweep or writing output to a second grid. The new value of each element is determined by a function of the values of neighboring elements.

Cellular automata provide a powerful model for computational simulation of complex systems. For instance, discrete event simulations using cellular automata have been used to model the spread of forest fires [36]. The space where the simulation occurs is modeled as discrete cells. Each cell can be in one of several states, and the cells evolve in discrete time steps based on their state and the state of neighboring cells.

Conway's Game of Life is a simple recreational example of cellular automata. Each cell has two states, alive or dead. The cells in a 2D Cartesian grid evolve from generation to generation according to simple rules based on their state and those of the adjacent 8 cells in their neighborhood. Figure 3.21 shows that cells are addressed by their row and column index. The eight neighbors of a cell at position (i, j) are cells at $(i - 1, j - 1)$, $(i - 1, j)$, $(i - 1, j + 1)$, $(i, j - 1)$, $(i, j + 1)$, $(i + 1, j - 1)$, $(i + 1, j)$, $(i + 1, j + 1)$. There are two options for dealing with cells on the boundary, which have only 3 or 5 neighbors. We can create special rules to deal with these cells, or we can use periodic boundary conditions. In the latter case the neighbors wrap around the boundary, so for instance the east neighbor of cell $(2, 4)$ in Figure 3.21 is the cell at $(2, 0)$.

A cell that is alive will only still be alive in the next generation if 2 or 3 neighbors are alive. A cell that is dead will be alive in the next generation only if it has three neighbors that are alive. Each cell is visited and its neighborhood examined to determine whether its state will change in the next generation. In the example in Figure 3.21 cell $(1, 3)$ dies because it only has one live neighbor. Cell $(3, 2)$ dies because it has four live neighbors. Cell $(4, 3)$ comes alive since it has 3 live neighbors.

Algorithm 3.5 uses periodic boundary conditions, where $a \bmod b$ gives the remainder of integer division a/b. Note that it alternates between two copies of the grid.

A decomposition of the grid into cells is natural since each cell can be updated independently. Each task contains the state of one cell. In order to update its cell, it needs to obtain the states of its neighbors from the tasks that own them. This introduces a dependence between tasks in the previous and the current generations, as shown in Figure 3.22.

Algorithm 3.5: Game of Life

Input: $n \times n$ grid of cells, each with a state of alive (1) or dead (0).
Output: evolution of grid for a given number of generations

Allocate empty *newGrid*
for *a number of generations* **do**
    Display *grid*
    foreach *cell at coordinate (i, j)* **do**
        updateGridCell(*grid, newGrid, i, j*)
    end
    swap references to *newGrid* and *grid*
end

Procedure updateGridCell(*grid, newGrid, i, j*)
    $sumAlive \leftarrow grid[(i - 1 + n) \bmod n, (j - 1 + n) \bmod n]$
               $+ grid[(i - 1 + n) \bmod n, j] + grid[(i - 1 + n) \bmod n, (j + 1) \bmod n]$
               $+ grid[i, (j - 1 + n) \bmod n] + grid[i, (j + 1) \bmod n]$
               $+ grid[(i + 1) \bmod n, (j - 1 + n) \bmod n] + grid[(i + 1) \bmod n, j]$
               $+ grid[(i + 1) \bmod n, (j + 1) \bmod n]$
    if $grid[i, j] = 0 \wedge sumAlive = 3$ **then**
        $newGrid[i, j] \leftarrow 1$
    else if $grid[i, j] = 1 \wedge (sumAlive = 2 \vee sumAlive = 3)$ **then**
        $newGrid[i, j] \leftarrow 1$
    else
        $newGrid[i, j] \leftarrow 0$
    end
end

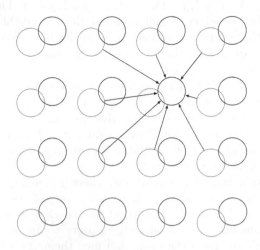

Figure 3.22: Game of Life task graph, showing dependencies of one task.

Matrix-vector multiplication

Matrix-vector multiplication is a fundamental operation that is used in many applications. It is also much easier to decompose than the multiplication of two matrices:

> **foreach** *row i of matrix A* **do**
> $b[i] \leftarrow 0$
>   **foreach** *column j of A* **do**
> $b[i] \leftarrow b[i] + A[i,j] * x[j]$
>   **end**
> **end**

We can think of matrix-vector multiplication as a series of inner products between b and a row of A (the inner loop above). This suggests a decomposition of A into rows, with each task assigned to compute an inner product of b with one of the rows of A, as we saw earlier in Figure 1.5. This produces an embarrassingly parallel decomposition, since these tasks are independent. However, to increase the number of tasks we can go further by recognizing that the inner product is a reduction operation. This leads us to decompose the matrix into its elements, and assign one task to the multiplication of one element of A and b. The results of tasks that were assigned elements from the same row of A are then combined using a parallel reduction.

As discussed above, an important guideline in designing a decomposition is to maximize the number of tasks. As we'll see in the next chapter, the next step is to coarsen the decomposition to a smaller number of tasks. For the Game of Life and matrix-vector multiplication we could consider assigning a task to a rectangular block of elements, rather than a single element.

Bottom-Up Dynamic Programming

Dynamic programming is an algorithmic technique to solve combinatorial optimization problems. The technique is similar to divide-and-conquer, but here there can be dependencies between subproblems. It finds the optimal solution to a problem as a function of optimal solution of subproblems. Dynamic programming algorithms are formulated as recurrence relations, and can be solved recursively. In practice they are usually solved in a bottom-up fashion, where the solution is built up from smaller to larger problems, and implementation involves computing the elements of a multidimensional table.

A simple example is the *subset sum* problem. Given a set of positive integers $\{s_1..s_n\}$, does there exist a subset whose sum is equal to a desired value S? This problem is solved using the following recurrence relation, where $F[i,j]$ is the answer to the question: does a subset of the first i integers sum to j:

$$F[i,j] = \begin{cases} F[i-1,j] \vee F[i-1,j-s_i] & \text{if } i > 0; \\ 1 & \text{if } j = 0; \\ 0 & \text{otherwise.} \end{cases}$$

For $F[i,j]$ to be true, this means that item s_i is either included in the subset that sums to j, or it is not. If it is included, then $F[i-1,j-s_i]$ must be true. If it is not included, then $F[i-1,j]$ must be true. If neither of those cases is true then $F[i,j]$ must be false. A base case is included for $j = 0$, for which an empty subset answers the question in the affirmative. This recurrence relation can be solved recursively, starting from $i = n$ and $j = S$, and expanding to two subclasses, then four, and so on, until an answer is found. Dynamic programming problems are normally solved by effectively going up the recursion

---

Algorithm 3.6: Subset sum

Input: Array $s[1..n]$ of n positive integers, target sum S
Output: returns 1 if a subset that sums to S exists, 0 otherwise
Data: Array $F[1..n, 0..S]$ initialized to 0

```
// subset sum always true for j = 0
```
for $i \leftarrow 1$ *to* n **do**
    $F[i, 0] \leftarrow 1$
end

$F[1, s[1]] \leftarrow 1$ `// subset of first integer summing to itself`

for $i \leftarrow 2$ *to* n **do**
    for $j \leftarrow 1$ *to* S **do**
        $F[i, j] \leftarrow F[i - 1, j]$
        if $j \geq s[i]$ **then**
            $F[i, j] \leftarrow F[i, j] \vee F[i - 1, j - s[i]]$
        end
    end
end
return $F[n, S]$

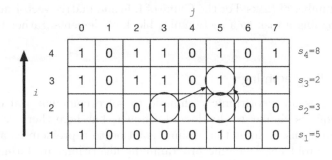

Figure 3.23: Subset sum solution for $s = \{5, 3, 2, 8\}$ and $S = 7$, showing the dependence of $F[3, 5]$ on $F[2, 5]$ and $F[2, 5 - s_3]$.

tree by filling in a table iteratively. For subset sum a 2D array of n rows and $S + 1$ columns is computed iteratively, starting from $i = 1$ and $j = 1$, as seen in Algorithm 3.6. Elements in the first column are precomputed to 1, and the only other nonzero entry in row $i = 1$ is for $j = s[1]$. The answer is given in the upper right value in the table ($i = n, j = S$): if it is equal to 1 the subset exists, otherwise it does not.

We can see that all values in a given row can be computed independently, since they only depend on values in the previous row. We can therefore decompose the table into $n \times S$ elements, and associate a task with each element. Each task in a row depend on two tasks from the previous row. It is interesting to note that one of the dependencies for each task is not uniform, but depends on the value of s_i, as seen in Figure 3.23. We have not included any optimizations of this algorithm, including stopping at the end of row i when $F[i, S] = 1$. The complete task graph for this example is shown in Figure 3.24.

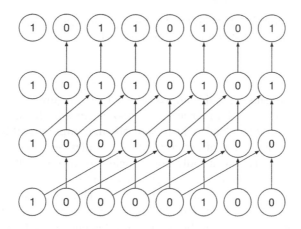

Figure 3.24: Task graph for subset sum example in Figure 3.23.

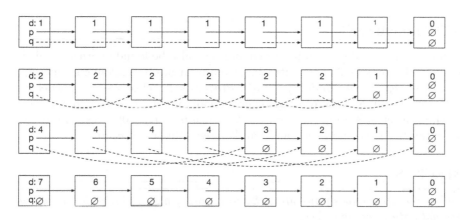

Figure 3.25: List ranking with pointer jumping.

Pointer Jumping

Data decomposition is not restricted to arrays. We can also decompose linked data structures, assigning a task to each element. Consider the problem of computing the distance of each node to the end of a linked list, which is called *list ranking*. Sequentially, this is easy. The length of the list n can be found by traversing it to the end. Then the list is traversed a second time, with the distance at the ith element being recorded as $n - i$. At first glance it would seem that this problem is inherently sequential, but we can solve it in parallel using a technique known as pointer jumping [74]. We decompose the list into its elements and assign a task to each element. Each task makes a copy q of the element's pointer and initializes its distance d to the end to 1, with the exception of the task representing the last element (whose pointer is null, and whose distance is 0). Then each task adds its value of d to that of the next element (using pointer q) and then replaces q with the next element's pointer q. This process continues for $\log n$ iterations and is illustrated in Figure 3.25. After the ith iteration the q pointers span 2^i elements. Tasks can work in parallel, though they must synchronize before every iteration. This allows the list ranking to be done in $O(\log n)$ steps if we assume that the assignment of tasks to elements has previously been done, which takes $O(n)$ time. Pointer jumping can be used to solve other problems, such as the prefix sum of values stored in a linked list, and can also be used with trees.

3.8 SUMMARY

Parallel algorithm design begins with decomposition, which can involve either decomposition into tasks or decomposition of data structures, or sometimes both. An important goal is to maximize the number of independent tasks so that the maximum possible parallelism can be explored. There's no consideration of the computational platform, such as the number of processors to be used. This exercise is usually not sufficient to produce efficient parallel programs. The computer's execution model(s) and the programming language model will favor particular decompositions. The number and size of tasks, that is, their granularity, needs to be chosen with the target computer in mind. In the next chapter we'll examine common parallel programming structures that can be used to implement algorithmic structures such as those described in this chapter.

3.9 FURTHER READING

The algorithmic structures discussed in this chapter can also be found in other parallel computing textbooks. Foster's PCAM (Partitioning, Communication, Agglomeration, Mapping) methodology in *Designing and Building Parallel Programs* [24] clearly separates decomposition (the "P" and "C" in PCAM) from agglomeration and mapping to a computer. Chapters 3 and 4 of Mattson et al.'s *Patterns for Parallel Programming* [50] gives a formal design pattern presentation of decomposition and parallel algorithmic structures.

3.10 EXERCISES

For all parallel decompositions draw the task graph, and describe the tasks and any additional data structures used.

3.1 The reduction of an array shown in Figure 3.6 has three steps, summing elements separated by a stride of 1, 2, and 4, respectively. Draw two possible ordering of sums with other permutations of strides 1, 2, and 4.

3.2 Design a parallel decomposition for the problem of finding the *second* largest value in an array.

3.3 Design a parallel decomposition for the problem of finding the index of the *first* occurrence of the value x in an array.

3.4 a. The k-means algorithm 3.1 assigns each vector to one of k cluster centers. Given such an assignment, design a parallel decomposition to find the size of each cluster.

   b. Design a parallel decomposition to write the vectors to a new array, in order of their cluster membership, making use of the results of exercise 3.4a.

3.5 Quicksort is another divide-and-conquer sorting algorithm:

```
// sort elements with index i ∈ [lower..upper)
Procedure Quicksort(a, lower, upper)
    if lower < upper − 1 then
        randomly select pivot element
        partition a: {i < mid | a[i] ≤ pivot}, a[mid] = pivot, {i > mid | a[i] > pivot}
        Quicksort(a, lower, mid + 1)
        Quicksort(a, mid + 1, upper)
    end
end
```

Where is most of the work done in this algorithm? Describe a naive parallel decomposition. Then design a parallel decomposition that improves on the naive approach.

3.6 Trace the execution of the pipelined merge sort of Algorithm 3.4, using the following array: $[3, 4, 6, 5, 8, 1, 2, 7]$.

3.7 A simple rhyming dictionary can be created from a text document by reversing each word, sorting the reversed words, and finally reversing each word. Describe a parallel algorithm to produce a rhyming dictionary for each of a large number of text files.

3.8 Draw the task graph for top-down recursive solution of the subset sum problem for which the bottom-up solution is give in Figure 3.23. Compare the number of tasks that are executed in both cases.

3.9 The Floyd-Warshall algorithm finds the length of the shortest path between all pairs of vertices in a directed graph.

Input: $n \times n$ adjacency matrix a, $a[i, j]$ giving weight of edge (i, j), ∞ if no edge.
Output: Matrix a, $a[i, j]$ giving weight of shortest path between vertices i and j.

```
for k ← 0 to n − 1 do
    for i ← 0 to n − 1 do
        for j ← 0 to n − 1 do
            a[i, j] = min(a[i, j], a[i, k] + a[k, j])
        end
    end
end
```

Design a parallel decomposition of this algorithm.

3.10 Describe an algorithm that uses pointer jumping to produce the prefix sum of a linked list of numbers.

3.11 Design a parallel decomposition to find the number of occurrences of a substring in another string.

CHAPTER 4

Parallel Program Structures

Good parallel algorithm design starts with task and/or data decomposition, without consideration of the eventual computing platform. This initial stage often reveals several options, and creates decompositions with as many independent tasks as possible. The next stage is to consider the programming platform and language that will be used. Each type of platform places particular constraints, and can guide the implementation in different ways. The computing power of desktops ranges from multicore processors, with a handful of cores, to GPUs with thousands of cores. These compute nodes can be rack-mounted and networked to form large clusters with thousands or even millions of cores. The assignment of tasks to processing units may be under the control of the programmer, or may be handled by runtime software, depending on the parallel programming model.

Chapter 2 introduced parallel machine models and the task graph execution model. Chapter 3 explored the construction of task graphs through a study of representative algorithmic structures. In this chapter we return to the three machine models (SIMD, shared and distributed memory), and examine how they influence parallel algorithm design and implementation. We explore some commonly used programming models and the program structures that they use, and recommend good design practice relevant to each.

Parallel program structures consist of tasks executing on processing elements (which we'll call cores). A primary concern when evaluating a parallel program is how well the work is balanced across the cores.

4.1 LOAD BALANCE

What is the execution time of a parallel program? This seems like an obvious question, but it's not quite as simple as for a sequential program. For both we can use the concept of *wall-clock time*, where we imagine starting a stopwatch as the program begins executing, and stopping the timer once the program completes. For a parallel program we start the timer as soon as the program is launched, even if not all the cores start executing at the same time. Then we need to wait for all the cores to finish execution before stopping the timer. Ideally all the cores would begin executing the program at the same time, and they would all complete at the same time.

Let's say we have a fixed amount of work that takes W steps to complete and that we measure execution time in terms of the number of steps. We also assume that all steps take the same amount of time. If p cores are used to do this work in parallel, then the best we can do is to assign an equal amount of work to each core, which results in an execution time of W/p (see Figure 4.1a). However, in general the execution time will be $T_p \geq W/p$. A

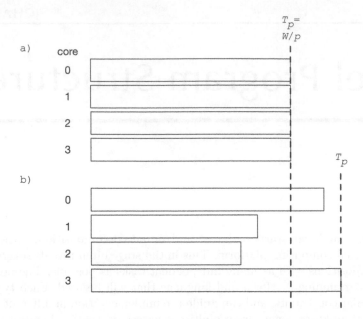

Figure 4.1: a) Perfectly balanced load and b) imbalanced load of work W on $p = 4$ cores.

commonly encountered reason for not reaching the lower bound of W/p is when the work assignment to the cores is unbalanced. In Figure 4.1b the workload is unbalanced, so the execution time is greater than W/p. In Figure 4.1 all the cores start at the same time. Load imbalance can also result when some cores delay the time they begin execution.

Load balance is an important concern when designing parallel algorithms, and we'll refer to it throughout this chapter. We'll look at performance more formally in Chapter 5, where we'll also explore the other factors that degrade performance.

4.2   SIMD: STRICTLY DATA PARALLEL

SIMD architectures have played a significant role in the history of parallel computers. They continue to be important because data parallel programming is very common in scientific computing and in graphics processing. Earlier in Section 2.1.1 we studied an example of array summation with SIMD instructions.

There are many such cases where the same operation is applied independently to consecutive elements of an array. This pattern has been called *strict data parallel* [44], where a single stream of instructions executes in lockstep on elements of arrays. This contrasts with data parallel programs executing on MIMD platforms: the multiple streams have identical instructions, but they don't operate in lockstep, relying instead on synchronization when all instruction streams must be at the same point. This latter type of programming is called *Single Program Multiple Data* (SPMD) and will be seen below when we examine programming on shared and distributed memory platforms.

SIMD programming can be done in several ways. A vectorizing compiler can produce vector instructions from a source program if the iterations of the loop are independent and the number of iterations is known in advance (see Section 2.1.1). Some programming languages feature array notation, which makes the compiler's job much easier. In this case arrays a and b can be summed by writing something like $c[0 : n - 1] \leftarrow a[0 : n - 1] + b[0 :$

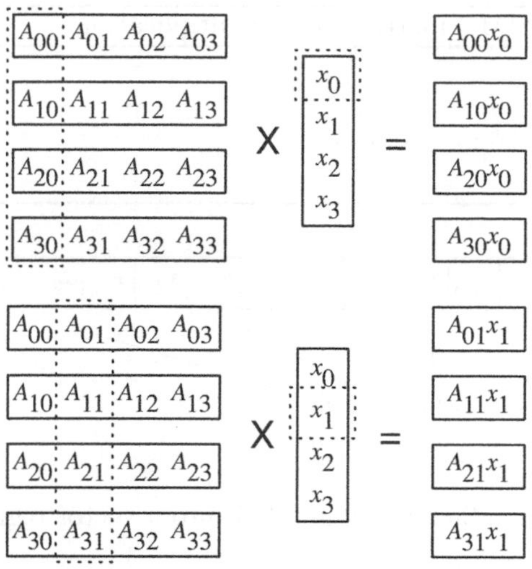

Figure 4.2: Two iterations of SIMD multiplication $b = Ax$, with a row-wise decomposition. Resulting values are accumulated in vector b.

$n-1]$, or even $c \leftarrow a + b$. This statement indicates explicitly that the operations occur independently on all elements. The compiler can then break down the sequence into chunks with a number of elements that depends on the number of bits in the data type and in the vector registers. The evocative name for this process is *strip mining*. For example, if we have arrays of 2^{20} double precision elements, and the platform is Intel's Xeon Phi co-processor, which features 512 bit vector registers, then the loop can be broken up into 2^{17} groups of 8, and the operations in each group are executed simultaneously.

There are other ways to describe strict data parallel loops, such the Fortran 95 `forall` loop construct. We will adapt the set notation from Blelloch and Maggs [9], as it is precise and easy to understand. Using this notation the array sum would be written $\{c[i] \leftarrow a[i] + b[i] : i \in [0..n)\}$, which represents SIMD parallel execution of $c[i] \leftarrow a[i] + b[i]$ for index i varying from 0 to $n-1$.

Row-Wise Matrix-Vector Multiplication

The parallel matrix-vector multiplication that was discussed in Chapter 3 resulted in a decomposition of the matrix into its elements. This introduces dependencies resulting from the inner product of each row of the matrix with the vector. If we consider the coarser decomposition into rows of the matrix, then each task can independently perform the inner product of a row of the matrix with the vector. In a shared or distributed memory implementation, each core independently iterates through the inner product of its row of A with x, as we saw in Figure 1.5 in Chapter 1. In the SIMD implementation of Algorithm 4.1, the inner products occur in lock-step: at each iteration of the inner product a column of A is multiplied by one of the elements of x and the result is accumulated in b, as illustrated in Figure 4.2. This is a row-wise decomposition because each inner product is done sequentially, but multiple inner products take place in parallel.

If A is stored in column-major order, then elements of a column will be contiguous in memory. This means that a column can likely be loaded in one transaction. If A is stored in row-major order, then the elements of a column are scattered in memory (Figure 4.3), and

---

Algorithm 4.1: row-wise SIMD matrix-vector multiplication

$\{b[i] \leftarrow 0 : i \in [0..n)\}$
for $j \leftarrow 0$ *to* $m - 1$ **do**
    $\{b[i] \leftarrow b[i] + A[i, j] * x[j] : i \in [0..n)\}$
end

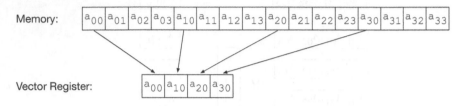

Figure 4.3: Loading a column of A into a vector register.

could take more transactions to load. While some SIMD machines support scattered memory loads and stores, in general better performance is obtained when the data is contiguous.

One could address this issue by coarsening the decomposition into columns of A, and perform the inner product of each row of A with b in parallel. This pattern is clearly a reduction, which we have treated in Chapter 3 as a binary tree of tasks.

Reduction

The binary tree of Figure 3.5 from Chapter 3 seems to be far from a strict data parallel pattern, until we view the pattern of operations on an array, as we did in Figure 3.6 (although here we will reduce into the first element). We can perform an in-place reduction of the elements of an array a in $\log n$ stages, where in stage k we perform $a[i] \leftarrow a[i] + a[i + 2^k]$. However, we don't want to perform this operation for each value of i. Instead, we only need to perform this operation for elements that satisfy $i \bmod 2^{k+1} = 0$. This means that we are not performing the same operation for each element, but only for a subset that meet a certain condition. Fortunately SIMD machines enable conditional execution, so we can perform an in-place reduction of the elements of an array a as:

    for $k \leftarrow 0$ *to* $\log n - 1$ **do**
        $j \leftarrow 2^k$
        $\{a[i] \leftarrow a[i] + a[i + j] : i \in [0..n) \mid i \bmod 2j = 0\}$
    end
    // `result in` $a[0]$

where we've added a conditional clause to our set notation which states that the operation will only be performed for index i if $i \bmod 2j = 0$.

This modification from the strict data parallel model is called *control divergence*. It can reduce performance, partly because of the overhead required to support conditional execution, but mainly due to the reduced number of operations that execute in parallel. Let's say the SIMD width can contain 4 elements of a. For the first stage of reduction ($k = 0$) only two additions could be done in parallel, and for the other stages the additions would all be serialized.

We could remove the conditional clause by observing that in iteration 0 the odd elements

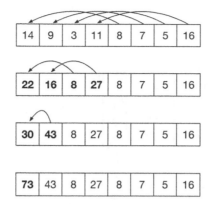

Figure 4.4: Divergence-free SIMD reduction.

are added to the even elements, in iteration 1 every fourth element is updated, and in general element $i * 2^{k+1}$ is replaced by its sum with element $i * 2^{k+1} + 2^k$:

for $k \leftarrow 0$ *to* $\log n - 1$ **do**
    $j \leftarrow 2^{k+1}$
    $\{a[i * j] \leftarrow a[i * j] + a[i * j + j/2] : i \in [0..n/j)\}$
end

While this removes the divergence, the memory access pattern is scattered, as the elements in each of the set of operands are separated by a stride of 2^k elements. This reduces performance, as we saw above with loading a column of a matrix. We can do better by recognizing that we are free to choose the pattern of additions. We can add each element in one half of the array with the corresponding element in the second half, and repeat recursively starting with the first half of the array, as in Figure 4.4 and Algorithm 4.2.

Algorithm 4.2: Divergence-free SIMD reduction

---

for $k \leftarrow \log n - 1$ *to* 0 **do**
    $j \leftarrow 2^k$
    $\{a[i] \leftarrow a[i] + a[i + j] : i \in [0..j)\}$
end
`// result in` $a[0]$

Now the operands occupy contiguous memory locations and there is no control divergence. At each iteration up to 2^k operations would be done in parallel, depending on the SIMD width. This algorithm can also be adapted to work with array sizes other than powers of two, by adding the last element to the previous one for each iteration of the k-loop where there are an odd number of operations for the SIMD loop.

Column-Wise Matrix-Vector Multiplication

In a column-wise decomposition each task multiplies a column of A with an element of x and accumulates the result in b. In each outer iteration of Algorithm 4.3 elements of a row of A are multiplied with the corresponding elements of x, as illustrated in Figure 4.5, and stored in array *temp*, then the elements of *temp* are summed using SIMD reduction.

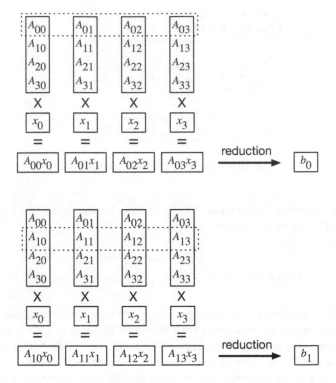

Figure 4.5: Two iterations of SIMD multiplication $b = Ax$, with a column-wise decomposition.

Algorithm 4.3: Column-wise SIMD matrix-vector multiplication

---

for $i \leftarrow 0$ *to* $n - 1$ **do**
    $\{temp[j] \leftarrow A[i,j] * x[j] : j \in [0..m)\}$
    for $k \leftarrow \log n - 1$ *to* 0 **do**
        $\{temp[j] \leftarrow temp[j] + temp[j + 2^k] : j \in [0..2^k)\}$
    end
    $b[i] \leftarrow temp[0]$
end

Subset Sum

Let's look now at another data parallel decomposition from Chapter 3, the subset sum problem, where the table was decomposed into its elements. The dependencies here are between the rows, while computation of elements in a row can be done independently. We can therefore coarsen the decomposition by agglomerating elements in each column, with a task for each element in a row. As the computation moves from row to row, each task computes the value of one element in the row, which depends on two elements in the previous row. A SIMD implementation gives, for the nested `for` loops of Algorithm 3.6:

> **for** $i \leftarrow 2$ *to* n **do**
> $\quad \{F[i,j] \leftarrow F[i-1,j] : j \in [1..S]\}$
> $\quad \{F[i,j] \leftarrow F[i,j] \vee F[i-1,j-s[i]] : j \in [1..S] \mid j \geq s[i]\}$
> **end**

The control divergence in the second statement isn't as severe as it was for our first implementation of reduction. It will only affect one chunk of operations that contains both $j < s[i]$ and $j \geq s[i]$. Chunks with $j < s[i]$ only would simply be skipped, and those with $j \geq s[i]$ only would be executed as if there were no divergence.

SIMD Guidelines

Exploitation of current SIMD-enabled processors does not require explicit parallel programming, but rather an understanding of the factors that affect performance, so that the statements that are doing data parallel computation can be structured properly. From the programmer's perspective, there are three important guidelines to consider when relying on strict data parallel execution on SIMD platforms:

1. *Watch out for dependencies*

2. *Avoid control divergence*

3. *Optimize memory access patterns*

If the compiler is relied on to vectorize loops, then the programmer must structure them in such a way that there are no dependencies between iterations. The other two guidelines need to be followed when optimizing performance.

The same guidelines apply to GPU programming. However, while the execution model is SIMD, the programming model is Single Program Multiple Data, as we'll see below.

4.3 FORK-JOIN

Strict data parallel parallel execution is fine-grained by nature. The more concurrent operations that can be identified the better, particularly for GPUs, where thousands of threads can execute concurrently. In general, the granularity of parallel decompositions needs to be assessed based on the characteristics of the computing platform. Tasks need to be assigned to processing units and their execution coordinated. If the tasks are too fine-grained then the overhead of managing them will dominate the execution time and hence the performance will be poor. The number of tasks can be reduced by agglomerating them into fewer tasks. However, it is not necessarily the case that the number of tasks should be the same as the number of cores. A larger number of tasks allows for load balancing, which may be required because of variability in task execution times or because of variability in performance of the cores.

Threads are used for shared memory parallel execution. Each thread has its own program counter and private memory region in its stack frame, and shares instructions and heap and global memory with other threads. Threads are scheduled to cores by the operating system and they may be migrated among cores, but they can also be pinned to particular cores through programmer intervention.

One implementation pattern that is well suited for divide-and-conquer task graphs is fork-join, since it implies a tree structure of tasks. Fork-join is supported in many parallel software environments, such as Pthreads, Cilk, OpenMP, and Java. In divide-and-conquer parallel programs tasks are recursively split into other tasks. When splitting occurs, new tasks are *forked*. This can also be called *spawning* a task. Before combining the results of its work with that of its children, the parent task must wait for its child tasks to complete. This can also be called *joining* or *synchronizing*.

Algorithm 4.4: Sequential recursive Estimation of π

```
// n = 2^k experiments
Procedure estimatePi(n)
    return recPi(n)*4/n
end

// returns sum of points in circle
Procedure recPi(n)
    if n = 1 then
        sum ← 0
        x ← pseudo-random number ∈ [−1, 1]
        y ← pseudo-random number ∈ [−1, 1]
        if x² + y² ≤ 1 then
            sum ← 1
        end
    else
        sum ← recPi(n/2) + recPi(n/2)
    end
    return sum
end
```

For a simple example consider the approximation of π using the Monte Carlo method, which was an example of an embarrassingly parallel decomposition in Chapter 3. We'll use a recursive solution so that we can use a parallel reduction to combine results. The n experiments can be divided in half, so that π is approximated by 4 times the sum of the number of points in the circle for each experiment divided by n (ratio of area of unit circle to 2×2 square is $\pi/4$). This process continues recursively until the base case where a point is generated. The sequential recursive algorithm, where we assume n is a power of 2, is shown in Algorithm 4.4.

This algorithm can be decomposed into tasks using a task graph in the form of a tree, where the points are generated at the leaves and the results are gathered up the tree, just as values are summed up a reduction tree. We can trace sequential execution mentally as it passes from the root down to the left-most leaf, where a point is generated, then the next leaf, then the results of those two leaves are combined, and so on. This corresponds to visualizing post-order traversal of a binary tree.

How can we mentally trace parallel execution? To do so we need to map tasks to threads that execute on different cores. Since we are assuming a shared memory environment then

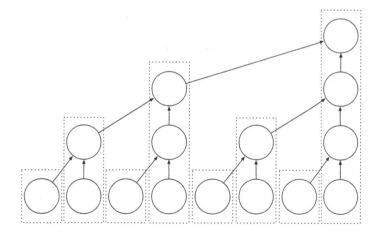

Figure 4.6: Assignment of 15 reduction tasks to 8 threads.

child tasks can share the results of their execution with their parents without any need for explicit communication. We can trace down and up the tree, with tasks on a level executing in parallel by threads on distinct cores. One could expect the tasks at each level to be created at roughly the same time. Once execution goes back up the tree then tasks may have to wait for both children to complete, but if all the tasks complete in roughly the same amount of time then task execution should mostly overlap in time at each level.

How many threads/cores should be used? If we assume for the moment that the number of cores is unbounded then we could assign a thread to each task in the graph. A parent thread would then spawn two child threads and then wait for them to complete before combining their results. This would yield $2n - 1$ threads, one for each task in the tree. Theres no reason, however, why a parent and its child tasks need to be executed by 3 different threads. The assignment of tasks to threads in Figure 4.6 shows that we can reduce the number of threads to n if each parent only creates a new thread for its left child, but executes its right child itself. You can see from the figure that we still have fully exploited the parallelism at each level of the tree.

In practice, of course, the number of cores isn't unbounded. We could still use n threads and let the operating system schedule them to cores, but the overhead of creating and scheduling so many threads would be too great. We could instead stop the spawning of threads at a high level in the recursion tree to limit the number of threads. It is good practice in general for divide-and-conquer programs to limit the depth of the tree by cutting off the recursion at an appropriate task size, and computing the leaves using a sequential algorithm.

Explicit assignment of tasks to threads can be problematic in practice. First, there is measurable overhead involved in creating and synchronizing threads. If tasks are too lightweight the overhead can affect performance. If we take the other extreme and only have coarse tasks then we could be limiting the achievable parallelism. Second, assignment of tasks to threads isn't always as obvious as in our example. Tasks may not take the same time to complete, which complicates their assignment to threads.

A better approach is to specify which tasks can be executed independently, and at what point parent tasks need to wait for child tasks to complete, and let runtime software take care of assigning tasks to threads. This is what is done for programming models that enable fork-join specification of tasks. The runtime creates a thread pool and allocates a work queue for each thread. As tasks are created they are placed in the queue of the parent thread. Threads can steal tasks from another thread to spread the work around.

For our fork-join algorithms we will use **spawn f()**, where f() is some procedure, to spawn tasks, and **sync** to wait for child tasks to complete, which is close to the syntax of Cilk. We could then simply change the recursive calls to recPi() to the following:

$sum1 \leftarrow$ **spawn** recPi$(n/2)$
$sum2 \leftarrow$ recPi$(n/2)$
sync
$sum \leftarrow sum1 + sum2$

It's important to emphasize that the **spawn** keyword doesn't imply the creation of a thread. This keyword only indicates that the execution of recPi$(n/2)$ can be done independently of the following statements up to the **sync** and it is left to the runtime to assign them to threads. Note that we didn't put a **spawn** in front of the second call of recPi(). As discussed above this would lead to an unnecessarily large number of threads (up to $2n + 1$) being used. This algorithm produces up to n independent tasks. There's good reason to further reduce this number, because of the overhead from scheduling so many tasks to a smaller number of threads, as n is likely orders of magnitudes larger than the number of threads. We don't want to decrease the number of tasks to equal the number of threads, as discussed above, but instead find a good compromise. This can be accomplished by providing a sequential cutoff, as shown in Algorithm 4.5.

Algorithm 4.5: Parallel fork-Join estimation of π

Procedure recPi(n)
    if $n <$ cutoff **then**
        $sum \leftarrow 0$
        for $i \leftarrow 1$ *to* n **do**
            $x \leftarrow$ pseudo-random number $\in [-1, 1]$
            $y \leftarrow$ pseudo-random number $\in [-1, 1]$
            if $x^2 + y^2 \leq 1$ **then**
                $sum \leftarrow sum + 1$
            end
        end
    else
        $sum1 \leftarrow$ **spawn** recPi$(n/2)$
        $sum2 \leftarrow$ recPi$(n/2)$
        sync
        $sum \leftarrow sum1 + sum2$
    end
    return sum
end

The value of the cutoff can be determined experimentally. There's no need to implement the leaf tasks recursively, and a loop is used instead. While the sequential computation of π can simply use a loop, the parallel version benefitted from a recursive divide-and-conquer decomposition to enable parallel accumulation of results. Parallel reduction using the fork-join pattern required only a minor changes to the recursive sequential algorithm. We could use the same approach to implement the decomposition of the k-means algorithm presented in the last chapter.

Merge Sort

Next let's apply the fork-join pattern to merge sort, a true divide-and-conquer algorithm. We'll not only spawn new tasks at each recursive level, but also perform the merge operation in parallel, as described in Chapter 3 in Section 3.5. We'll use a sequential cutoff, and the base case can use any efficient $O(n \log n)$ sorting algorithm. Similarly, when the sequential cutoff is reached in the parallel merge the regular sequential merge is done. Recall that in order to split the merge in two we find the median *mid* of the longest of the two sorted subarrays and do a binary search of the other subarray to find the index that splits it into elements less than and greater than *mid*.

The `parMergeSort` procedure in Algorithm 4.6 is a straightforward modification of the sequential merge sort of Algorithm 3.3 to include the **spawn** and **sync** keywords. The parallel merge discussed in Section 3.5 is quite different from the sequential version, but is still very simple, with the slight complication to ensure that the index of the median element is found for the larger subarray. The `parMerge` procedure needs 2 indices for each of the lower and upper subarrays, since they won't always be contiguous. We can trace the execution starting from the rapid recursive calls and spawning of tasks until the base case is reached, at which point approximately $n/$`cutoff` sequential sorts will take place, which are scheduled to threads by the fork-join runtime. This will enable good overlap in time for the threads executing the sequential sorts. The pairs of subarrays are then merged in parallel with each merge involving tasks executing in parallel (see Figures 3.13 and 3.16).

Fork-Join Guidelines

As the parallel merge algorithm makes clear, the fork-join pattern greatly simplifies implementation of divide-and-conquer algorithms, as there's no need to figure out how to allocate tasks to threads in order to balance the load of the cores that run the threads. There are three guidelines that should be followed:

1. *Don't create more independent tasks than necessary*

2. *Use a sequential cutoff to limit the depth of the recursion*

3. *Avoid unnecessary allocation of memory*

4. *Be careful if shared read/write access to data required*
   Warning: *may be hidden inside functions*

In the case of merge sort, the first guideline means that we only used **spawn** to create a single additional task at each level of recursion. The third guideline means that divide and merge operations should not take extra overhead. For the parallel merge we could have allocated temporary arrays to store merged subarrays, but this would have unnecessarily added to the execution time. Merging alternatively between preallocated arrays avoids the need for temporary arrays. The last guideline is to be very careful about sharing data between tasks. In the parallel merge, tasks worked on disjoint sections of the arrays, so there was no need to guard against simultaneous use of shared data. Mutual exclusion needs to be used if there is read/write access to shared data, as will be discussed below in Sections 4.4.1–4.4.3. There could very well be such a problem lurking in Algorithm 4.5 in the function used to generate pseudo-random numbers (see Section 4.4.3).

Algorithm 4.6: Fork-join parallel merge sort.

Input: array a of length n.
Output: array b, containing array a sorted. Array a overwritten.

```
arrayCopy(a, b)  // copy array a to b
parMergeSort(a, 0, n, b)
```

```
// sort elements with index i ∈ [lower..upper)
```
Procedure parMergeSort(a, $lower$, $upper$, b)

    if $(upper - lower) <$ cutoff **then**

        sequentialSort(a, $lower$, $upper$, b)

    else

        $mid \leftarrow \lfloor (upper + lower)/2 \rfloor$

        spawn parMergeSort(b, $lower$, mid, a)

        parMergeSort(b, mid, $upper$, a)

        sync

        parMerge(a, $lower$, mid, mid, $upper$, b, $lower$)

        return

    end

end

```
// parallel merge of sorted sub-arrays i ∈ [low1..up1) and i ∈ [low2..up2)
//    of a into b starting at index start
```
Procedure parMerge(a, $low1$, $up1$, $low2$, $up2$, b, $start$)

    $k1 \leftarrow up1 - low1$

    $k2 \leftarrow up2 - low2$

    if $k1 + k2 <$ cutoff **then**

        sequentialMerge(a, $low1$, $up1$, $low2$, $up2$, b, $start$)

    else

        if $k1 \geq k2$ **then**

            $mid1 \leftarrow \lfloor (low1 + up1 - 1)/2 \rfloor$

            // $mid2$: first index in $[low2, up2)$ such that $a[index] > a[mid1]$

            $mid2 \leftarrow$ binarySearch(a, $low2$, $up2$, $mid1$)

        else

            $mid2 \leftarrow \lfloor (low2 + up2 - 1)/2 \rfloor)$

            $mid1 \leftarrow$ binarySearch(a, $low1$, $up1$, $mid2$) -1

            $mid2 \leftarrow mid2 + 1$

        end

        spawn parMerge(a, $low1$, $mid1 + 1$, $low2$, $mid2$, b, $start$)

        parMerge(a, $mid1 + 1$, $up1$, $mid2$, $up2$, b, $start + mid1 - low1 + 1 + mid2 - low2$)

        sync

    end

end

One might be tempted to apply fork-join to everything, including data parallel decompositions, but it's better to use it for implementing divide-and-conquer decompositions. Fork-join doesn't simplify the implementation of data parallel algorithms and there is a price to pay in the overhead of task scheduling. Even for divide-and-conquer algorithms it may be worth considering iterative implementations, as we'll see later in this chapter.

4.4 PARALLEL LOOPS AND SYNCHRONIZATION

One of the simplest ways to express shared memory parallelism is to declare a loop to be executable in parallel:

parallel for $i \leftarrow 0$ *to* $n - 1$ **do**
$\quad c[i] = a[i] + b[i]$
end

where the body of the loop can be run independently for each iteration. Parallel loops can be used to express part of a parallel algorithm or they can also be used to incrementally parallelize an existing program. The latter use case has resulted in the overwhelming success of the OpenMP standard API, which began its history by enabling parallel loops. The iterations of a parallel **for** loop are scheduled by the compiler and runtime for execution by threads. There is an implicit *barrier* at the end of the loop, which all threads must reach before execution continues after the loop.

Parallel loops are particularly applicable to embarrassingly parallel and data parallel algorithms that contain tasks operating independently on distinct elements of a data structure. Incremental parallelization of loops in existing sequential code requires careful dependency analysis and as a result can lead to refactoring to remove dependencies. We won't address this subject here, which is well treated by the literature on parallelizing compilers and programming models with parallel loops.

Matrix-vector multiplication provides a simple example. We've already seen that a simple way to parallelize this operation is to independently perform the inner product of the rows of the matrix with the vector:

parallel for each *row i of matrix A* **do**
$\quad b[i] \leftarrow 0$
$\quad$ **foreach** *column j of A* **do**
$\quad\quad b[i] \leftarrow b[i] + A[i, j] * x[j]$
$\quad$ **end**
end

Loop Schedules

What the **parallel for** notation doesn't reveal is how the iterations of the outer loop will be executed in parallel by threads. One could use a divide-and-conquer method, whereby the loop is recursively split in two until a threshold is reached and let a fork-join framework schedule tasks. Or one could assign the iterations to threads in contiguous chunks, that is each thread would execute iterations $\lfloor id * n/nt \rfloor$ to $\lfloor (id + 1) * n/nt \rfloor - 1$, where there are nt threads each with an $id \in [0..nt)$. For example if $n = 2048$ and $nt = 5$, threads would execute iterations as follows:

thread 0: $i \leftarrow 0$ to 408
thread 1: $i \leftarrow 409$ to 818
thread 2: $i \leftarrow 819$ to 1227
thread 3: $i \leftarrow 1228$ to 1637
thread 4: $i \leftarrow 1638$ to 2047

Since 2048 mod 5 = 3 the iterations can't be divided equally among threads, so 3 threads ($id = 1, 3, 4$) do one extra iteration. Alternatively, iterations could be assigned in smaller chunks in round-robin fashion. For chunk size 1 iterations 0 through 9 would be run by threads 0, 1, 2, 3, 4, 0, 1, 2, 3, 4, respectively.

The previous two schedules are *static*, in that they can be produced at compile time. *Dynamic* schedules are also possible, where the runtime assigns small chunks of iterations to threads, and once threads complete their work they are assigned another chunk. This approach is an example of the master-worker pattern, which we will see below.

Fractal

Let's contrast these different ways of scheduling iterations using the fractal generation example from Chapter 3. Recall that we generate the color of a pixel, whose coordinates form the real and imaginary parts of the complex number c, based on how many iterations of the recurrence $z_{i+1} = z_i^\alpha + c$ it takes to diverge. The initial condition z_0 is $(0, 0)$ for $\alpha > 0$ and $(1, 1)$ for $\alpha < 0$. The sequential algorithm for a $n \times n$ pixel image is given in Algorithm 4.7.

We can agglomerate tasks in each row, and allocate the computation of rows to threads by replacing the outermost `for` with a `parallel for`. If more tasks are desired we could agglomerate fewer than n tasks into rectangular regions of the image. Clearly the number of iterations of the innermost loop will not be uniform across pixels, as we can see by looking at a typical fractal in Figure 3.4. Recall that white pixels take the maximum number of iterations (don't diverge) whereas darker pixels diverge within a few iterations. Rows in the top half of the image have more white pixels than those in the bottom half and hence take longer to compute.

If a fork-join framework is used, with sufficiently more tasks than threads, then the runtime should be able to balance the work of all threads. The first static schedule, with contiguous chunks, will suffer from some threads having more work than others, which will increase the overall execution time. The round-robin schedule should produce better results than the other static schedule as it is more likely that the generation of white regions of the image will be done by more threads. Finally, the dynamic schedule would be expected to produce similar performance as the fork-join approach.

Subset Sum

In the previous two examples the work of each thread was completely independent, so no synchronization was required. Let's look again at the subset sum problem. We can start with S tasks as in the SIMD implementation, one for each element in a row:

Algorithm 4.7: Generation of fractal $z = z^\alpha + c$

```
// Image coordinates: lower left (xmin, ymin) to upper right
   (xmin + len, ymin + len)
```
`// `$xmin = ymin = -1.5$` and `$len = 3$` for full image`

Input: α, n, $xmin$, $ymin$, len

Output: $n \times n$ pixel fractal

Data: niter `// max iterations`

threshold `// threshold for divergence`

$ax \leftarrow len/n$
$ymax \leftarrow ymin + len$
for $i \leftarrow 0$ *to* $n - 1$ **do**
    $cx \leftarrow ax * i + xmin$
    for $j \leftarrow 0$ *to* $n - 1$ **do**
        $cy \leftarrow ymax - ax * j$
        $c \leftarrow (cx, cy)$
        if $\alpha > 0$ **then**
            $z \leftarrow (0, 0)$
        else
            $z \leftarrow (1, 1)$
        end
        for $k \leftarrow 1$ *to* niter **do**
            if $|z| <$ threshold **then**
                $z \leftarrow z^\alpha + c$
                $kount[i, j] \leftarrow k$
            else
                break `// exit inner loop`
            end
        end
    end
end

> **for** $i \leftarrow 2$ *to* n **do**
>   **parallel for** $j \leftarrow 1$ *to* S **do**
> $F[i, j] \leftarrow F[i - 1, j]$
>   **if** $j \geq s[i]$ **then**
> $F[i, j] \leftarrow F[i, j] \vee F[i - 1, j - s[i]]$
>   **end**
>   **end**
> **end**

Here the threads do not operate in lockstep, as in the SIMD implementation. Each thread gets a chunk of iterations to perform. Iterations 1 to $s[i] - 1$ of the parallel loop do not execute the body of the conditional statement, so they will take less time. This means the thread schedule should be chosen such that these iterations don't all go to one thread, to avoid load imbalance. The implicit barrier at the end of a **parallel for** loop is needed here because of the dependence between computation of rows (see Figure 3.24).

So far it seems that loop parallelization is a trivial matter. However, we need to be very careful about shared and private variables. Eliminating data races allows us to use sequential consistency to reason about our program and ensure correctness, as we saw in Chapter 2.

4.4.1 Shared and Private Variables

Recall that threads can store local variables in their stack frame and access variables in heap and global memory that are shared with other threads. It's important for the programmer to ensure that variables that should be private to a thread are either declared inside the loop body or are annotated as private by the programming API. Clearly, the iteration counters of the parallel loop and all nested loops inside need to be private for each thread. For matrix-vector multiplication, where the amount of work is the same for each row, one might expect each thread to execute in lockstep, and hence variable j could be shared. However, threads aren't guaranteed to be synchronized, for reasons such as differences in memory access times and competition for CPU resources with other processes. This means that different threads could easily be at different iterations of the inner loop at any given time. Loop counters aren't the only concern. Implementation of the parallel fractal program would also need to make sure that variables c, cx, cy, and z were private.

4.4.2 Synchronization

The edges in a task graph of a parallel decomposition represent real data dependencies between tasks. We have to take these dependencies into account in a shared memory implementation. For instance, the dependence between selected tasks in adjacent rows in the parallel subset sum above introduces a data race. Each iteration reads from $F[i - 1, j]$ and $F[i - 1, j - s[i]]$, which are memory locations that were written to in the previous iteration. The simplest way to eliminate this data race is to put a barrier after each iteration, which we've said is implicit in the **parallel for**. A barrier is a synchronization point that all threads must reach before crossing, and memory updates resulting from the execution of threads before the barrier are visible to all threads.

While a barrier is a commonly used synchronization operation, there are also situations where only local synchronization between a subset of threads is desired. For instance, use of the barrier between iterations of the subset sum implies dependence between all tasks at adjacent iterations, whereas there are only two inter-iteration dependencies (see Figure 3.24).

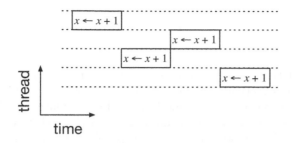

Figure 4.7: Atomic increment of x by four threads.

One can get better performance by directly specifying the dependencies between tasks, as we'll see below in Section 4.5. Other cases where point-to-point synchronization is needed is when one or more threads need to wait until one thread produces a value, in so-called producer-consumer problems.

Even if we carefully specify which variables are shared and private and use barriers to ensure values that need to be read have already been written, we can still miss some data races. We saw in Chapter 2 that we can be misled into thinking that a high level operation, such as incrementing a variable, is atomic. To take up the estimation of π again, we could try doing it with a parallel loop:

```
// Danger, produces indeterminate result!
Procedure iterPi(n)
    sum ← 0
    parallel for i ← 0 to n − 1 do
        x ← pseudo-random number ∈ [−1, 1]
        y ← pseudo-random number ∈ [−1, 1]
        if x² + y² ≤ 1 then
            sum ← sum + 1
        end
    end
    return sum * 4/n
end
```

Clearly, x and y need to be private variables and sum needs to be shared. This program will produce incorrect results that will vary each time it is run, because of the data race involving reading and writing to sum. We can ensure that $sum \leftarrow sum + 1$ is executed atomically, and hence get rid of the data race, by enclosing it in a *critical section*:

```
begin critical
sum ← sum + 1
end critical
```

A critical section ensures that the execution of the enclosed statements by multiple threads does not overlap in time, and thus makes the statements atomic, as illustrated in Figure 4.7. This property of a critical section is called *mutual exclusion*.

Parallel language models offer different ways to express a critical section. With OpenMP one can simply declare a block to be critical using a pragma annotation. The most common way to express a critical section is to use a lock API:

```
lock()
sum ← sum + 1
unlock()
```

A lock enforces the mutual exclusion property of a critical section. In Java the **synchronized** keyword is used to make a method critical, so that it can only be invoked by one thread at a time. The body of knowledge concerning locks and other mutual exclusion constructs is vast, and beyond the scope of this book. Locks have to be used very carefully in order to protect all accesses to shared variables and avoid problems such as deadlock. Locks can also introduce significant overhead in execution time. Explicit use of locks can be avoided by using libraries of concurrent data structures, such as linked lists and hash tables, which encapsulate safe access to shared data. Another alternative is to use lower level atomic instructions to implement critical sections.

Compare and Swap

One of the most common atomic instructions is compare-and-swap (CAS), which enables the atomic update of a variable in a single hardware instruction.

Algorithm 4.8: Compare and Swap

---

atomic Procedure cas($\&x$, old, new)

    if $x = old$ **then**

        $x ← new$

        return true

    else

        return false

    end

end

In Algorithm 4.8 $\&x$ uses C notation to refer to the address of variable x. The way a CAS is used is that a thread reads the current value of x then calls cas() to update x with the new value, which only succeeds if x still has the current value. The CAS will fail if another thread changed the value of x before the cas() was called. The CAS itself is performed atomically in hardware, so only one thread at a time can execute it. We can thus increment sum by repeatedly calling cas() until it returns true:

repeat

    $old ← sum$

    $new ← sum + 1$

until cas($\&sum$, old, new) = true

The use of CAS doesn't always involve looping until the condition is met. For example, consider a simple algorithm to remove duplicates of an array a of elements that have a limited range of nonnegative values. Algorithm 4.9 uses an array t of size equal to the range of values to record the values that are found in a. The updates of t are made atomic using a CAS. The update only needs to change the value of an element of t from 0 to 1, so no action is needed if the update fails due to the element already having been set to 1. Once the parallel loop is complete a sequential loop writes the unique values to a.

Algorithm 4.9: Remove duplicates

Input: array a of n nonnegative integers in the range with maximum value $high$.
Output: array a with duplicates removed, with k values.
Data: array t with $m = high + 1$ elements, initialized to 0.

parallel for $i \leftarrow 0$ *to* $n - 1$ **do**
    cas($\&t[a[i]]$, 0, 1)
end
$k \leftarrow 0$
for $i \leftarrow 0$ *to* $m - 1$ **do**
    if $t[i] = 1$ **then**
        $a[k] \leftarrow i$
        $k \leftarrow k + 1$
    end
end

The ABA Problem

There is an important limitation to the use of a CAS, in that it is only based on the value of a memory location. That's fine if this is just considered to be a value, but what if it refers to a pointer? Consider using CAS to implement a function to atomically pop a stack implemented as a linked list, where top points to the first node and $top \to next$ points to the second node:

Procedure pop()
    repeat
        $old \leftarrow top$
        $new \leftarrow (top \to next)$
    until cas($\&top$, old, new) = true
    return old
end

Consider the following sequence of operations, and a stack initially consisting of $top \to A \to B \to C$:

thread 0: $old \leftarrow top$
thread 0: $new \leftarrow (top \to next)$
thread 1: $a \leftarrow$ pop() //$top \to B \to C$
thread 1: $b \leftarrow$ pop() //$top \to C$
thread 1: push(a) //$top \to A \to C$
thread 0: cas($\&top$, old, new)//$top \to B$

After thread 0 makes old point to A and new point to B, thread 1 pops two nodes and pushes back the first node, leaving top still pointing to node A. Then thread 0 is able to successfully issue the cas() and top now points to B. Of course this is incorrect as top does not point to the top of the stack but instead points to B, which doesn't exist anymore! The problem is that the fact that the value of the pointer to the top of the stack was unchanged after the intervention of thread 1, although the stack itself clearly had changed. Fixing this problem involves keeping track not only of the value but also of the number of changes, which can be implemented using spare bits in the pointer or a CAS that can operate atomically on two values.

Alternatives to Critical Sections

One of the drawbacks of using critical sections is that they reduce performance by serializing the enclosed operations, as seen in Figure 4.7. A better way to resolve the race condition in $sum \leftarrow sum + 1$ would be to have each thread accumulate sums in a private variable and then have one thread add them up, as in Algorithm 4.10.

Algorithm 4.10: Reduction without critical section.

> **Procedure** iterPi(n)
>> $sum \leftarrow 0$
>> // i, id, x, y private
>> **parallel for** $i \leftarrow 0$ *to* $n - 1$ **do**
>>> $id \leftarrow$ getThreadID()
>>> $x \leftarrow$ pseudo-random number $\in [-1, 1]$
>>> $y \leftarrow$ pseudo-random number $\in [-1, 1]$
>>> **if** $x^2 + y^2 \leq 1$ **then**
>>>> $psum[id] \leftarrow psum[id] + 1$
>>> **end**
>> **end**
>> **for** $i \leftarrow 0$ *to* $nt - 1$ **do**
>>> $sum \leftarrow sum + psum[i]$
>> **end**
>> **return** $sum * 4/n$
> **end**

Here each thread accumulates its sum in a separate element of shared array $psum$. After the parallel loop is complete the elements of $psum$ are summed sequentially.

The reduction pattern is so common that shared memory APIs and programming languages offer ways to declare reduction variables. These reduction variables are private to each thread, but are summed together after the loop completes, preserving the semantics of the sequential loop.

As with this example it is good practice to avoid shared data structures when possible and instead use private data and gather results. For instance, in the histogram example at the beginning of Chapter 3 the second decomposition had each task putting one value in a histogram and then all histograms were merged. Implementation using a shared memory loop would allocate a chunk of iterations to each thread. This could be implemented using a single shared array and with each thread making atomic updates. Given that there are only 16 bins the likelihood of threads having to wait to update a bin would be high. Better performance would be likely if each thread worked on its private copy of the histogram and at the end a reduction was used.

4.4.3 Thread Safety

While you may be very careful to avoid data races in your code, you also need to be aware of data races in functions written by someone else, as these functions may not have been designed to be called concurrently by multiple threads. Functions that can be called by multiple threads without any data races occurring are called *thread-safe* functions. Such functions are also called *re-entrant*, as one thread can safely enter a function while another thread is already executing it.

All library functions should be checked to see if they are thread safe. Functions that do I/O should be thread safe. Implementation of `recPi()` and `iterPi()` must ensure that a thread-safe pseudo-random number generator is used.

Guidelines for Parallel Loops

Parallel loops are particularly applicable to implementing shared memory data parallel algorithms. Concerns include eliminating data races, through appropriate specification of private and shared variables and use of synchronization and ensuring load balance through experimentation with loop schedules:

1. *Eliminate data races*

2. *Load balance with loop schedules*

4.5   TASKS WITH DEPENDENCIES

In Chapter 3 we saw that parallel decompositions can be viewed as task graphs, even for data decompositions. For implementation on shared memory computers the fork-join and parallel loop patterns can handle many algorithmic patterns. The former exploits the tree structure of recursive decompositions and the latter exploits the independent updates to a data structure that arise in data decompositions. To see why one might want other options, consider the synchronization each pattern offers. In fork-join the synchronization is local and hierarchical: a task spawns other tasks, and then waits for their completion. Parallel loops use global synchronization, as all threads wait at a barrier at the end of the loop. What if one wishes to synchronize arbitrary tasks?

The subset sum algorithm provides an example of how one can avoid global synchronization. Consider the parallel loop implementation, where each thread computes a chunk of the elements of a row. All threads have to wait until the last thread has completed its work before proceeding to the next row. One can see from the task graph in Figure 3.24, however, that each task only has dependencies on two tasks in the previous row. This means a task could start executing once the two tasks it depends on have finished, even if other tasks in the previous rows have not finished.

Parallel language models that enable specification of task dependencies simplify the implementation of parallel algorithms without sacrificing performance. Armed with task dependencies, runtime software can schedule tasks to cores of a parallel computer more efficiently than a programmer can. All that the programmer needs to do is to specify the tasks and their dependencies. The dependencies can be specified *indirectly*, through data dependencies, or *directly*. For the former one could specify the input and output data dependencies of each task. One task might write a value to x and another task might read the value of x, which would result in a dependence between the two tasks. One can instead directly specify the dependencies between tasks. The independent data-driven approach has the advantage that it can be used for both shared memory and distributed memory environments. Given the data dependencies, the runtime can take care of the necessary data communication. This is the approach taken by the Legion programming model [5].

The direct specification of dependencies is more straightforward, but is applicable only to shared memory environments because of the lack of information on data dependencies. We'll

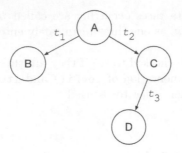

Figure 4.8: Task graph for Example 4.1.

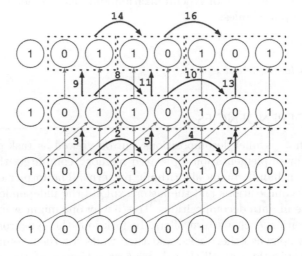

Figure 4.9: Blocked wavefront execution of subset sum example from Figure 3.24. Diagonal dependencies between rows have been replaced with horizontal dependencies between neighboring blocks in the same row. Dependencies are labeled with integers (see Algorithm 4.11). Tasks in the first row and first column are computed sequentially.

focus on this approach, as it naturally allows implementation of task graph decompositions, and it gives a taste of programming with task dependencies. It is supported by some parallel language models, including OpenMP. We'll add an **in** and an **out** clause to the **spawn** keyword. Edges in the task graph are identified by an integer, which can be a constant or an expression. For example:

Example 4.1

```
spawn out(t1, t2) A()
spawn in(t1) B()
spawn in(t2) out(t3) C()
spawn in(t3) D()
```

where t_1 through t_3 are distinct labels and the associated task graph is shown in Figure 4.8. Tasks B and C can begin executing in parallel as soon as task A has completed. Task D can start once task C is complete, regardless of whether task B is finished. As with fork-join frameworks, tasks are scheduled dynamically based on their dependencies by runtime software. Note that there is no need for a `sync` keyword, as all synchronization is implicitly specified by the dependencies.

Starting from the original decomposition of the subset sum problem, one could spawn a task for each element in the table, indicating the `in` and `out` dependencies. The problem is that this would create too many tasks, which would overwhelm the runtime scheduling, so it's better to agglomerate tasks in a row into contiguous chunks. Unfortunately, this complicates figuring out the task dependencies that depend on the data (diagonal arrows in Figure 3.24).

To simplify our solution, we can change the dependencies in a manner that still provides the benefit of avoiding a barrier after each row. Instead of computing a row of tasks in parallel, we can compute them in a diagonal wavefront starting from the first task in the bottom left and moving toward the top right task, as shown in Figure 4.9. In the original decomposition the tasks in each row were independent. In this wavefront decomposition tasks along each diagonal are independent. We can accomplish this by changing the dependency associated with $F[i-1, j-s[i]]$ to instead point to the block immediately preceding the current block in the same row [27]. This ensures that before a block is computed all the elements have been computed in the current row and the row below with column index less than those in the current block.

To label the dependencies we assign an even integer to horizontal dependencies and an odd integer to vertical ones. This gives us Algorithm 4.11. The initial `spawn out(2, 3)` starts the wavefront, then we proceed to spawn tasks in row $i = 2$ (the first row is computed sequentially), where each task has only one inward dependency, then continue with the remaining rows. Each block corresponds to elements with index $(j-1)\lfloor S/nB \rfloor + 1$ to $j\lfloor S/nB \rfloor$ ($j = [1..nB]$) in a row, where nB is the number of blocks per row. To help in computing the dependence labels we assign an even integer to each block, starting with $iB = 2 * ((i-2) * nB + 1)$ for the first block in each row, with the following blocks in a row labelled in increments of 2. This gives horizontal and vertical dependencies of iB and $iB + 1$ respectively. The inward dependencies of a block are computed by counting back twice the number of blocks to the source block, and subtracting the result from the block's horizontal and vertical dependencies, giving $iB - 2$ (the block immediately to the left) and $iB + 1 - 2 * nB$ (the block immediately below).

Once the block with outward dependencies $(2, 3)$ has been computed, then execution can begin for blocks with outward dependencies $(4, 5)$ and $(8, 9)$, and the wavefront continues until the top rightmost block is computed. Note the special case for the first column of tasks that have no horizontal inward dependencies. Note also that no global synchronization is required. This approach has been shown to result in better performance than a parallel loop implementation [27].

Guidelines for Programming with Tasks and Dependencies

This is still an active area of research, so it's difficult to distill some guidelines. As with other programming models, it's important to consider its suitability for the problem at hand. Data parallel algorithms with few dependencies are more suitable for parallel loops. Recursive algorithms can be handled well using the fork-join model. Problems with complex dependencies, on the other hand, are good candidates for the task-based programming model.

Algorithm 4.11: Parallel subset sum with task dependencies

Input: Array $s[1..n]$ of n positive integers, target sum S, number nB of blocks per row

Output: returns 1 if a subset that sums to S exists, 0 otherwise

Data: Array $F[1..n, 0..S]$ initialized to 0

for $i \leftarrow 1$ *to* n **do**
    $F[i, 0] \leftarrow 1$
end

$F[1, s[1]] \leftarrow 1$

spawn out$(2, 3)$ calcRowChunk$(2, 1, \lfloor S/nB \rfloor)$
for $j \leftarrow 2$ *to* nB **do**
    spawn in$(2(j-1))$ out$(2j, 2j+1)$
                calcRowChunk$(2, \lfloor(j-1) * S/nB\rfloor + 1, \lfloor j * S/nB \rfloor)$
end
for $i \leftarrow 3$ *to* n **do**
    $iB \leftarrow 2 * (i - 2) * nB + 2$
    spawn in$(iB - 2 * nB + 1)$ out$(iB, iB+1)$ calcRowChunk$(i, 1, \lfloor S/nB \rfloor)$
    for $j \leftarrow 2$ *to* nB **do**
        $iB \leftarrow iB + 2$
        spawn in$(iB - 2, iB - 2 * nB + 1)$ out$(iB, iB+1)$
                    calcRowChunk$(i, \lfloor(j-1) * S/nB\rfloor + 1, \lfloor j * S/nB \rfloor)$
    end
end
return $F[n, S]$

Procedure calcRowChunk$(i, j1, j2)$
    for $j \leftarrow j1$ *to* $j2$ **do**
        $F[i, j] \leftarrow F[i-1, j]$
        if $j \geq s[i]$ **then**
            $F[i, j] \leftarrow F[i, j] \vee F[i-1, j - s[i]]$
        end
    end
end

4.6   SINGLE PROGRAM MULTIPLE DATA

The models discussed so far allow straightforward implementation of shared memory algorithms, and they leave the mapping of tasks to cores to the compiler and/or the runtime software. Also, it is understood that code blocks that aren't in the scope of a parallel directive will be executed sequentially by a single thread. The Single Program Multiple Data (SPMD) model requires the programmer to specify the mapping of tasks to threads. This increases the complexity of program development, but higher performance than loop parallelism can be achieved because of the greater control over work assignment that can encompass a larger part of the program (not just loops). It has been mainly applied to distributed memory programming using the Message Passing Interface, although it was actually first proposed in the late 1980s for a shared memory machine, IBM's Research Parallel Processing Prototype [14], and is still a viable model for shared memory programming today. It is also the basis for parallel programming on GPUs.

The idea is simple. All threads execute the same program, and the thread id is used to determine the work it will carry out. For parallel loops, for instance, each thread is explicitly assigned iterations based on its id. SPMD has mainly been applied to data parallel programming, where each thread executes the same instructions on different data, but it is also applicable to more general parallel programming, as different threads can execute different instructions. Shared memory SPMD programming can make use of shared and private data, and coordination of thread execution is accomplished using synchronization. By contrast, in distributed memory SPMD programming there is no shared data, so sharing and coordination is accomplished by sending messages. We'll discuss distributed memory SPMD programming in Section 4.8.

We have to be careful to distinguish between variables that are private to a thread and those that are shared by all threads, as we did with parallel loops. We will assume in our shared memory SPMD pseudocode that all the variables are private by default, and declare those that are shared.

For example, for matrix-vector multiplication we can explicitly assign iterations to threads:

shared a, b, c
$istart \leftarrow \lfloor id * nrows/nt \rfloor$
$iend \leftarrow \lfloor (id + 1) * nrows/nt \rfloor - 1$
for $i \leftarrow istart$ *to* $iend$ **do**
    $c[i] \leftarrow 0$
    foreach *column* j *of* a **do**
        $c[i] \leftarrow c[i] + a[i, j] * b[j]$
    end
end

where nt is the number of threads. Notice that for $nt = 1$ the result is exactly the sequential algorithm, and for $nt > 1$ each thread executes the same instructions, but for different iterations of the outer loop.

For the fractal calculation (Algorithm 4.7) we can apply a round-robin assignment of iterations to threads, in Algorithm 4.12. The rows are divided into chunks of size $chunk$. Thread id computes chunks beginning at rows $id * chunk$, $id * chunk + nt * chunk$, $id * chunk + 2 * nt * chunk$, etc.

These two examples illustrate how to manually assign iterations of a loop to threads, but they could have been implemented more easily using the parallel loop model. Another SPMD example can be found in a non-recursive algorithm for merge sort. The idea is to use data decomposition rather than a recursive divide-and-conquer approach. The array is

---

Algorithm 4.12: Round-robin SPMD parallel fractal generation

```
// chunk << n is chunk size of round-robin assignment of rows to
   threads
shared kount // array of pixel values is shared
istart ← id * chunk
iend ← (id + 1) * chunk − 1
ax ← len/n
ymax ← ymin + len
while istart < n do
    for i ← istart to iend do
        cx ← ax * i + xmin
        for j ← 0 to n − 1 do
            cy ← ymax − ax * j
            c ← (cx, cy)
            if α > 0 then
                z ← (0, 0)
            else
                z ← (1, 1)
            end
            for k ← 1 to niter do
                if |z| < threshold then
                    z ← zᵅ + c
                    kount[i, j] ← k
                else
                    break // exit inner loop
                end
            end
        end
    end
    istart ← istart + nt * chunk
    iend ← min(istart + chunk − 1, n − 1)
end
```

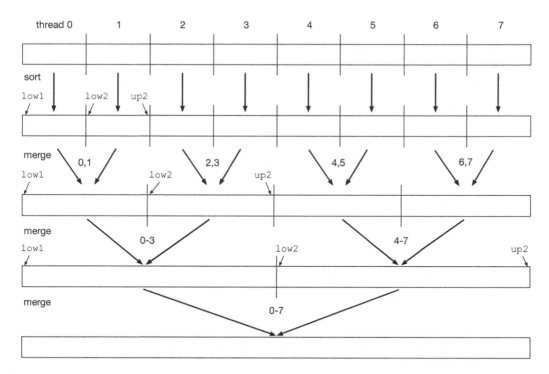

Figure 4.10: SPMD merge sort with 8 threads, showing indexes of subarrays for first merge at each level and indexes of threads involved in each operation (Algorithm 4.13).

divided into nt contiguous chunks and each thread sorts its chunk using a sequential sorting algorithm. Before the merging begins all threads must be finished completing their sort. This requires barrier synchronization. Then the nt chunks are merged in $\log nt$ stages.

Instead of the recursive divide-and-conquer merge, here we decompose the two subarrays into contiguous chunks in such a way that merges can be done in parallel. We can decompose the first subarray into contiguous chunks of approximately equal size, but we can't simply do the same with the other subarray. We can do this using binary search, as we did in the divide-and-conquer approach of Algorithm 4.6. A barrier is required after every stage to ensure that all subarrays have been merged before they are used in the next stage.

Algorithm 4.13 has two procedures, spmdMergeSort for sequential sorting and a loop of $\log nt$ iterations to perform the merges, and spmdMerge to perform the merges. For ease of exposition it assumes that the number of threads nt is a power of 2 and $n \bmod nt = 0$.

We have more work to do as a programmer than in the recursive algorithm, as we have to assign work to each thread and have to figure out the appropriate subarray indexes. It's important to emphasize that in SPMD programs all threads execute the same program, and their id determines the data they work on. First each thread sorts a contiguous chunk, then all nt threads do parallel merges in $\log nt$ stages. In each stage $i \in [1.. \log nt]$ $nt/2^i$ pairs of arrays are merged in parallel, and each merge itself is done in parallel using $nmt = 2^i$ threads, as shown in Figure 4.10.

In procedure spmdMergeSort the merges, which are done by calling spmdMerge, are indexed by thread id as $idc = \lfloor id/nmt \rfloor * nmt$. For example, in the first level of merges ($nmt = 2$) in Figure 4.10 threads 0 and 1 perform merge 0, threads 2 and 3 perform merge 1, and so on. The indexes that define the two subarrays are determined based on the merge

---

Algorithm 4.13: SPMD merge sort of array of size n by nt threads. $nt = 2^i$ and $n \bmod nt = 0$

Procedure spmdMergeSort(a, b)
    shared a, b
    $lower \leftarrow \lfloor id * n/nt \rfloor$
    $upper \leftarrow \lfloor (id+1) * n/nt \rfloor$
    sequentialSort(a, $lower$, $upper$, b) // sort for $i \in [lower, upper)$
    barrier()
    $nmt \leftarrow 1$
    for $i \leftarrow 1$ *to* $\log nt$ **do**
        Swap references to a and b
        $chunk \leftarrow nmt * n/nt$
        $nmt \leftarrow nmt * 2$
        $idc \leftarrow \lfloor id/nmt \rfloor * nmt$ // threads $id = idc$ to $idc + nmt - 1$ do each
            merge
        $low1 \leftarrow idc * n/nt$
        $low2 \leftarrow low1 + chunk$
        $up2 \leftarrow low2 + chunk - 1$
        // nmt threads merge $a[low1, low2)$ with $a[low2, up2]$ into b starting
            at index $low1$
        spmdMerge(a, $low1$, $low2$, $up2$, b, nmt)
        barrier()
    end
end

Procedure spmdMerge(a, $low1$, $low2$, $up2$, b, nmt)
    shared a, b
    $idm \leftarrow id \bmod nmt$
    $lowX \leftarrow \lfloor idm * n/(2 * nt) \rfloor + low1$
    $upX \leftarrow \lfloor (idm+1) * n/(2 * nt) \rfloor + low1 - 1$
    if $idm \neq 0$ **then**
        // first index $\in [low2, up2]$ such that $a[index] > a[lowX - 1]$
        $lowY := binarySearch(a, low2, up2 + 1, lowX - 1)$
    else
        $lowY \leftarrow low2$
    end
    if $idm < nmt - 1$ **then**
        $upY \leftarrow$ binarySearch(a, $lowY$, $up2 + 1$, upX) $- 1$
    else
        $upY \leftarrow up2$
    end
    $start \leftarrow lowX + lowY - low2$
    // merges $a[lowX, upX]$ with $a[lowY, upY]$ into b starting at index $start$
    sequentialMerge(a, $lowX$, $upX + 1$, $lowY$, $upY + 1$, b, $start$)
end

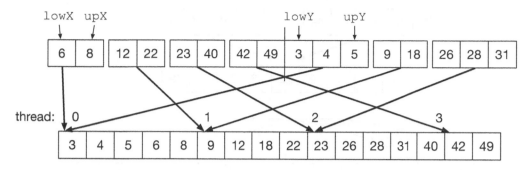

Figure 4.11: SPMD merge with 4 threads, showing indexes of subarrays for merge done by thread 0.

index idc and the size of each subarray, which is given by $(nmt/2) * n/nt$. The merges take place alternatively between arrays a and b, with a barrier at the end of each iteration.

Procedure **spmdMerge** merges $a[low1, low2)$ with $a[low2, up2]$ into b starting at index $low1$. An example is shown in Figure 4.11 for a merge with 4 threads. The subarray $a[low1, low2)$ is divided into nmt chunks and each thread computes the lower and upper indexes $lowX$ and upX of its chunk. The starting and ending indexes $lowY$ and upY of the other subarray are determined using two binary searches. The lower index $lowY$ is given by the index of the first element in $a[low2, up2]$ that is greater than $a[lowX - 1]$, and the upper index is given by the last index in $a[low2, up2]$ that is less than or equal to $a[upX]$, with border cases at the beginning and end of the array. Note that the first binary search for $lowY$ is done over the entire second subarray, while the second binary search for upY can ignore the elements before the lower index. Now each thread can perform a sequential merge for its subarrays.

Note that we could get away with having each of $nt - 1$ threads do a single binary search and write the results to a shared temporary array, so all threads could determine the boundary indexes of the chunks in the second subarray. We can do even better by not searching the entire second subarray. This is accomplished by splitting both subarrays into equal sized chunks and determining the chunk to search in the second subarray corresponding to each chunk in the first subarray.

Consider merging arrays x and y. To keep it simple we'll assume that all threads merge an equal size chunk of size c of array x. A subset is created of each array consisting of every cth element. Subset S_x contains the elements that split x into chunks, and the same is done for S_y. The element at index i of S_x is the last element of chunk i of array x. Instead of searching through all the elements of y to find the index that splits y into elements greater and less than $x[i]$, as we did in Algorithm 4.13, we need to find which chunk in y to search. To accomplish this we merge subsets S_x and S_y, producing S. For a given element $S_x[i]$ the index of the chunk of y that we need to search is given by $j - i$, where j is the index of $S_x[i]$ in S.

For example, consider the parallel merge from Figure 4.11. Figure 4.12 illustrates the procedure to find the elements that split the second subarray into chunks. The arrays S_x and S_y, containing every second element of subarrays x and y, are merged into S. To find the splitters in y we need to find which chunk of y to search in. For the last element of the first chunk in x, 8, the index of the chunk to search in y is given by its rank in S minus its rank in S_x, which is $1 - 0 = 1$. Searching in chunk 1 in y for the first element greater than 8 gives 9. For the last element of the second chunk in x, 22, the same procedure gives the

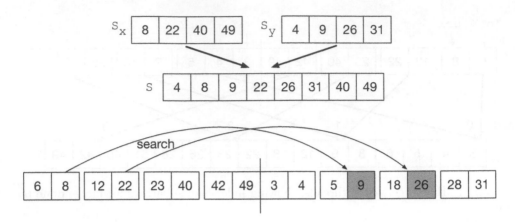

Figure 4.12: Searching for elements of subarray y (in gray) that delineate chunks matching chunks in subarray x. Results in chunks of second subarray y shown in Figure 4.11.

index $3 - 1 = 2$, so we search chunk 2 to get the value 26. The last two searches go past the end of array y. As a result, we have the splitters 9 and 26, which produce the chunks of subarray y shown in Figure 4.11. Of course, in practice the chunks will be much larger, so we'd be doing a much smaller merge in order to determine the parameters for a parallel merge. This algorithm was used to implement a merge sort on an Nvidia GPU [56].

If your head is spinning, you'll have seen one problem with SPMD programming: it can be complex to determine the parameters of the work assigned to each thread. Another problem is the need for global synchronization, which is not required for the fork-join algorithm. One of its advantages arises from the control of work assigned to threads, which can increase the locality of references. Recall the discussion in Chapter 2 of the hierarchical structure of multiprocessors. The more threads can keep working on their own data, or on data that is shared with a small group, the better the performance will be. For instance, in the SPMD parallel merge, the same threads that sorted adjacent subarrays also cooperate to merge them. In the fork-join merge since threads aren't assigned work explicitly, but instead steal work from each other dynamically, they could process tasks involving several arbitrary subarrays.

GPU Programming

Shared memory SPMD programming has been particularly successful on GPUs, as it trades increased programming complexity for very high performance. While the programming model is SPMD the execution model is SIMT, that is, small groups (warps) of threads are executed in SIMD fashion (see Section 2.1.1). Another feature that GPU programming brings is organizing threads into groups, and allowing threads in a group to share explicitly managed high speed memory and to perform barrier synchronization with other threads in the group. This hierarchical organization deals with the challenge of managing the thousands of threads that can be active simultaneously in a GPU. Very high performance is achievable if threads make good use of data shared by their group and if local synchronization in a group is favored over global synchronization. The significantly large number of threads highlights the importance of creating a decomposition with as many concurrent tasks as possible.

As an example, take the SIMD reduction Algorithm 4.2. It can be adapted in SPMD

Algorithm 4.14: Reduction on GPU

Procedure reduceGPU(a)

    // set group size and number of groups and allocate memory on GPU

    ...

    reduceToGroup(a, c)

    $sum \leftarrow 0$

    foreach *item in c* **do**

        $sum \leftarrow sum + item$

    end

    return *sum*

end

Procedure reduceToGroup(a, c)

    shared b // shared among threads in group

    $tid \leftarrow$ getThreadID()

    $gid \leftarrow$ getGroupID()

    $grpSz \leftarrow$ getGroupSize()

    $nt \leftarrow grpSz *$ getNumGroups()

    $id \leftarrow gid * grpSz + tid$

    $istart \leftarrow \lfloor id * n/nt \rfloor$

    $iend \leftarrow \lfloor (id + 1) * n/nt \rfloor - 1$

    $psum \leftarrow 0$

    for $i \leftarrow istart$ *to iend* **do**

        $psum \leftarrow psum + a[i]$

    end

    $b[tid] \leftarrow psum$

    syncGroup()

    for $k \leftarrow \log grpSz - 1$ *to* 0 **do**

        $j \leftarrow 2^k$

        if $tid < j$ **then**

            $b[tid] \leftarrow b[tid] + b[tid + j]$

        end

        syncGroup()

    end

    $c[gid] \leftarrow b[0]$

end

---

fashion for a GPU, in Algorithm 4.14. Tasks are agglomerated so that an array of length n is reduced with $nt < n$ threads. The `reduceGPU()` procedure sets the group size and number of groups, and runs on the CPU. The `reduceToGroup()` procedure runs on the GPU and produces an array of partial sums, one for each group. In this case `reduceGPU()` sums this array sequentially, but it could easily use `reduceToGroup()` to sum array c. This process could continue iteratively until the length of the partial sum array was short enough to make a sequential sum more efficient.

In `reduceToGroup()` each thread determines its index in array a based on its index in the group and the group's index. Each thread then sums a contiguous chunk of the array and stores the result in array b, which is shared by threads in a group. Then threads in each group synchronize with a barrier, and they then use the same SIMD algorithm as Algorithm 4.2 to sum their values, with the addition of a barrier to make sure all threads are synchronized after each iteration. Structuring the loop this away avoids the problem of thread divergence in a SIMT GPU.

Note that we have not included data transfer between GPU and CPU, which is required if the former is an attached co-processor. Depending on the programming language, this could be done explicitly via function calls or implicitly via parameter passing.

SPMD Guidelines

SPMD has been the dominant programming model for many years, particularly for distributed memory programming with message passing. It also forms the basis for GPU programming. A program is written for a single thread or process, whose rank is used to determine the data that will be used and the instructions to be executed. The assignment of work and data to threads/processes is done manually. Communication either occurs explicitly through messages or implicitly through shared memory. This section has given a taste of the complexity of SPMD programming. What it offers in return is potentially very high performance. The factors that affect performance will be discussed in Chapter 5. In addition, there are extensive implementation guidelines from the literature on MPI and CUDA/OpenCL programming.

In this section we've seen SPMD in a shared memory context, and the examples have used static work assignment. In Section 4.8 we'll see how SPMD is used in message passing programming. First, in the next section we'll look at a very common pattern that dynamically assigns work to tasks.

4.7 MASTER-WORKER

In our discussion of parallel loops we referred to the different ways loop iterations could be scheduled. One of the schedules is a dynamic assignment of work, where each thread processes a chunk then obtains another chunk. This schedule is particularly appropriate where the execution time of tasks is unpredictable, as in the fractal computation. A dynamic schedule can be implemented using SPMD programming with the master-worker pattern. This is a common pattern for dynamic load balancing of independent tasks. In its shared memory version the master sets up a queue of tasks and the workers dequeue the tasks and execute them. The master takes care of gathering the results if necessary. Care must be taken in handling access to the shared queue to avoid conflict between a thread adding to the queue and another thread removing from the queue. In the master-worker pattern only the master thread adds tasks to the queue. The master either populates the queue in advance, if the tasks are known, or adds them during execution if the tasks are created dynamically,

Algorithm 4.15: Master-worker fractal generation.

shared *kount, chunkCount*
if $id = 0$ **then**
    $chunkCount \leftarrow nt * chunk$
end
$istart \leftarrow id * chunk$
$iend \leftarrow (id + 1) * chunk - 1$
`barrier()`
while $istart \leq n - 1$ **do**
    for $i \leftarrow istart$ to $iend$ **do**
        `...// see sequential Algorithm 4.7`
    end
    begin critical
        $istart \leftarrow chunkCount$
        $chunkCount \leftarrow chunkCount + chunk$
    end critical
    $iend \leftarrow \min(istart + chunk - 1, n - 1)$
end

as occurs in divide-and-conquer algorithms for example. Race conditions can also occur if multiple threads attempt to remove tasks from the queue simultaneously, therefore the dequeuing operation must be made atomic using a mutual exclusion mechanism.

The master-worker pattern can be used to implement the fractal calculation using dynamic load balancing. A task consists of the calculation of a chunk of rows of the fractal. Here an explicit queue isn't needed; instead a shared counter keeps track of the progress of the computation. In Algorithm 4.15 a critical section is used so that only one thread at a time can identify its next chunk and update the chunk counter. The only work the master needs to do is to initialize the chunk counter, therefore it also acts as a worker. Barrier synchronization before the while loop ensures that the master has initialized the chunk counter before threads enter the loop.

The master-worker pattern can also be used to implement the fork-join algorithmic pattern, but a work-stealing pattern is more commonly used. In work-stealing each thread has a double ended queue, or deque. Each thread pushes and pops tasks from the head of its own deque. If there are no more tasks the thread takes a task from the rear of another randomly selected deque [46]. Work-stealing is not normally implemented by the programmer, but is used as the underlying mechanism by languages such as Java and Cilk.

Master-Worker Guidelines

The master-worker programming structure is very easy to describe and to use. One parameter that needs to be explored is the grain size of the tasks allocated to workers, such as *chunk* in Algorithm 4.15. If the tasks are too fine grained, workers may become idle waiting for their next work assignment. Making tasks too coarse can result in poor load balancing.

This structure has a more severe limitation in that it does not scale well as the number of workers increases. Once a threshold number of workers is reached, determined by the size of messages required and the complexity of the processing required per worker, the benefit of dynamic load balancing is negated by the overhead. One alternative uses multiple master tasks organized hierarchically, which spreads out the load balancing overhead [6].

4.8 DISTRIBUTED MEMORY PROGRAMMING

Our discussion of shared memory programming rarely mentioned any details of the underlying hardware platform, other than the effect of cache memory, the desirability of maintaining locality of memory references, and mutual exclusion techniques. In a multiprocessor the programmer's task is eased by having access to a shared address space. The hardware platform is more evident to the programmer in distributed memory computing. A distributed memory parallel computer consists of a number of processors interconnected by a switched network. While it is possible to create a single global address space, either in hardware or in software, most of these parallel computers are assemblies of separate computers with their own operating system and memory.

The details of the network topology are not of direct concern to the parallel programmer. Algorithms can be designed for particular topologies, such as a torus or hypercube, but they suffer from a lack of portability. The implementation details of common communication patterns, such as broadcast and reduce, are best left to libraries. What is important is the hierarchical organization of processing units, the memory capacity at each level and the latency and bandwidth between levels. The agglomeration of tasks and mapping to processing units must take into consideration these parameters, as they can vary widely. For example, communication between nodes in a high speed local network will have much lower latency than between nodes geographically distributed in a wide area network. We'll consider the impact of these parameters on performance in the next chapter.

An important concern for the programmer is how to distribute data structures across the compute nodes. Partitioned Global Address Space (PGAS) parallel programming models offer a shared memory abstraction for distributed memory computers [12, 55]. In PGAS languages the programmer specifies the distribution of arrays, but can access them as if they were in shared memory. Accesses to array sections that are not local will take longer than those to a local section.

The dominant programming paradigm is message passing. Message passing programming is typically done using the SPMD model, and dependencies between tasks are handled by explicit messages. The Message Passing Interface (MPI) standard has been the most successful parallel programming API for parallel computers. MPI has allowed highly scalable parallel algorithms to be implemented on large supercomputers, with portable programs that run on any parallel computer. Message passing offers high performance at the cost of considerable programming complexity. Parallel programming frameworks have been created to reduce this programming effort. We'll look at the most popular distributed computing pattern, MapReduce, which is used by several frameworks.

In shared memory computing the basic execution units were threads. In distributed memory programming the basic units are processes. As in the case of threads there can be more than one process active per core, but in practice there is usually only one process per core. In the distributed programming model there is no memory shared between processes, so they must access remote data through the exchange of messages.

4.8.1 Distributed Arrays

Data decomposition most often, but not exclusively, involves arrays. In the finest decomposition, each task is concerned with updating a single element, but as we have seen, these tasks are typically agglomerated into larger sections. We have examined agglomeration into one contiguous block per thread, and into a larger number of blocks that are then allocated cyclically to threads. Agglomeration works similarly in programming distributed memory computers. Consider matrix-vector multiplication. The agglomeration for shared memory

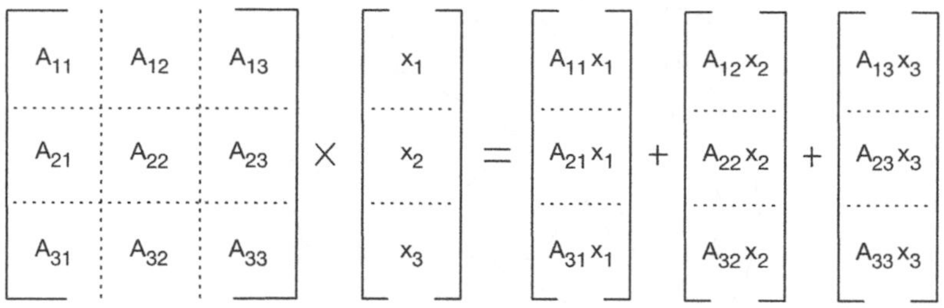

$$\begin{bmatrix} A_{11} & A_{12} & A_{13} \\ A_{21} & A_{22} & A_{23} \\ A_{31} & A_{32} & A_{33} \end{bmatrix} \times \begin{bmatrix} x_1 \\ x_2 \\ x_3 \end{bmatrix} = \begin{bmatrix} A_{11}x_1 \\ A_{21}x_1 \\ A_{31}x_1 \end{bmatrix} + \begin{bmatrix} A_{12}x_2 \\ A_{22}x_2 \\ A_{32}x_2 \end{bmatrix} + \begin{bmatrix} A_{13}x_3 \\ A_{23}x_3 \\ A_{33}x_3 \end{bmatrix}$$

Figure 4.13: 2D decomposition of matrix-vector multiplication.

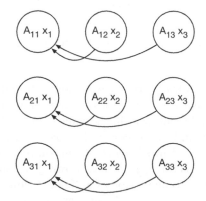

Figure 4.14: Task graph for matrix-vector multiplication with 2D decomposition.

had each task computing the inner product of one or more rows with the vector. This will also work for a distributed memory agglomeration. The vector can't simply be shared, however, but must be replicated across all tasks since it is required in its entirety for each inner product.

Another important factor in distributed memory agglomeration is how the initial and final states are constrained. Production parallel programs typically involve multiple stages. In practice matrix-vector multiplication isn't done in isolation, but is one component of an algorithm, so the agglomeration has to be consistent with the other stages. It may be, for instance, that a row-wise partition is consistent with the previous stage but the vector comes scattered among tasks. In this case the vector would have to be gathered by each process, and after the multiplication was complete the resulting vector would have to be scattered across all processes.

The agglomerations of data decompositions we've seen so far have only been along one dimension. For multidimensional arrays, agglomeration across more than one dimension is worth considering. Consider a two-dimensional agglomeration of the tasks in matrix-vector multiplication into d^2 tasks, by partitioning the matrix into $n/d \times n/d$ rectangular blocks $A_{i,j}$ and the vector into n/d blocks x_j. Figure 4.13 shows a 3×3 decomposition. It helps to visualize the tasks in a virtual $d \times d$ grid, as shown in Figure 4.14. To simplify the presentation we're assuming that matrix A is square. Each task has its own block $A_{i,j}$ of matrix A, and tasks in the same column have replicas of a block x_j of vector x. After completing the $A_{i,j} * x_j$ multiplication, each task has a partial result yp_j. To produce the final result, tasks in the same row perform a sum reduction of their vectors yp_j, with the results being held by tasks in the first column.

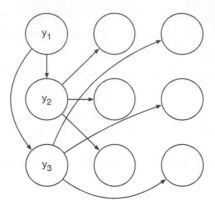

Figure 4.15: Redistribution of result of matrix-vector multiplication with 2D decomposition.

Our block-wise matrix-vector multiplication leaves the result matrix scattered among the tasks in the first column. To be consistent with the distribution of vector x, blocks of the result vector needs to be replicated in all tasks in the same column. Keeping the distribution of y consistent with that of x would be necessary, for instance, if multiple matrix-vector multiplications were being performed in sequence. To accomplish this redistribution each task in the first column sends its vector y_j to a task in row j, as shown in Figure 4.15.

It's also possible to construct a cyclic version of a 2D partition. The 2D array is divided into a larger number of blocks than tasks, and each task is assigned multiple blocks in a cyclic fashion, just as is done in a 1D decomposition.

It's evident that the lack of shared memory adds a layer of complexity in the distribution of data to tasks. The communication required for this distribution increases the execution time, so needs to be minimized. The relative merits of the different agglomerations will be discussed in Chapter 5.

4.8.2 Message Passing

SPMD is the most common distributed parallel programming model, where each task has the same program but the data and control path can differ based on its rank. Tasks can only operate on their local data. Dependencies between tasks are handled by sending messages. The message protocol is typically two-sided, where the sender and receiver each make a function call to transmit a message. The message header carries information such as the ranks of the sender and receiver, and a message tag to distinguish different types of messages. The message body is in the form of an array of data. Data to be sent may have to be packed by the sender if it does not reside in a contiguous region of memory. The overall communication pattern of a message-passing application can be classified into two types. Communication can be local, which only involves two tasks at a time, or it can be global, where all or a subset of tasks are involved. An example of the latter is the reduction seen in the matrix-vector example.

Local Communication

Send and receive operations can be *synchronous* or *asynchronous*, and *blocking* or *nonblocking*. A *synchronous* send operation only completes once the receiver has received the message. An *asynchronous blocking* send operation completes as soon the data region be-

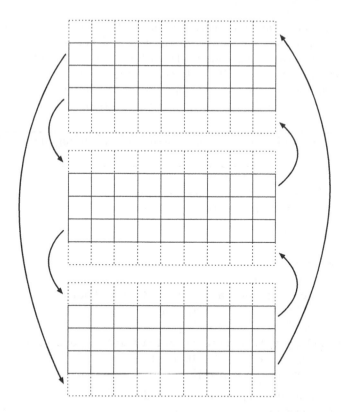

Figure 4.16: 1D decomposition of Game of Life grid with ghost cells, showing communication pattern of rows for 3 tasks.

ing sent can safely be overwritten, whereas an *asynchronous nonblocking* send completes immediately. An asynchronous blocking send is different from a synchronous send since the message is copied to a buffer, so it can be rewritten, before it has been transmitted. Two tasks can exchange messages by each making a nonblocking send followed by a receive:

> **if** $id = 0$ **then**
>> nonblocking send data to 1
>> receive data from 1
> **else**
>> nonblocking send data to 0
>> receive data from 0
> **end**

Both sends will return immediately and both receives will return once the message has been stored. It is important to observe that *deadlock* would occur if the sends above were synchronous, since then each task would be waiting for the other task to receive its message. The exchange of data between pairs of tasks is so common that MPI provides a function that performs the exchange safely.

An example of a local communication pattern is found in Conway's Game of Life. Let's assume that a contiguous block of rows is assigned per task. Each task can apply the sequential algorithm to its block of cells, but cells on the upper and lower boundaries of the block depend on values from neighboring tasks. This means that before each generation each task must exchange boundary information with neighboring tasks. To accommodate this, each task has an extra boundary layer on the top and bottom of its block, known as *ghost* cells (see Figure 4.16). Each task sends its boundary cells to the ghost cells of its neighbors, as shown in Algorithm 4.16. Note that if the boundary conditions were not periodic each task would have to check if the upper or lower boundaries on its sub-grid lie on the boundary of the overall grid, and take appropriate action.

Algorithm 4.16: Game of Life with Message Passing, not including initialization and display of grid

---

// each task has $(n/p + 2) \times n$ arrays $grid$ and $newGrid$
Input: $n \times n$ grid of cells, each with a state of alive (1) or dead (0).
Output: evolution of grid for a given number of generations

$nbDown \leftarrow (id + 1) \bmod p$
$nbUp \leftarrow (id - 1 + p) \bmod p$
$m \leftarrow n/p$ // Assume $n \bmod p = 0$
for *a number of generations* **do**
    // nonblocking send of boundary values to neighbors
    nonblocking send $grid[m, 0..n - 1]$ to $nbDown$
    nonblocking send $grid[1, 0..n - 1]$ to $nbUp$
    // receive boundary values from neighbors into ghost elements
    receive from $nbDown$ into $grid[m + 1, 0..n - 1]$
    receive from $nbUp$ into $grid[0, 0..n - 1]$

    foreach *cell at coordinate* $(i, j) \in (1..m, 0..n)$ **do**
        updateGridCell($grid, newGrid, i, j$)
    end
    swap references to $newGrid$ and $grid$
end

Global Communication: Scatter and Gather

Algorithm 4.16 has omitted two important elements: how the grid is initialized and how results are collected and displayed. If we assume that the initial state is in a file, then one task could read the grid and distribute the initial values to other tasks. This operation, called a *scatter*, can be implemented using a binary tree in a manner similar to a reduction, as we'll see in the next chapter. Similarly, one task needs to gather results from other tasks periodically so they can be displayed. A *gather* operation works the same way as a scatter, only with the directions reversed.

In practice, implementing collective operations is complex, as the network topology and other machine characteristics need to be taken into account. Programmers shouldn't implement their own collective communication operations, but instead should rely on libraries that are implemented appropriately for each platform.

Four common global communication patterns are illustrated in Figure 4.17. In our simplified API, for `scatter` and `broadcast` the first and second parameters indicate the *source* task rank and the *source* data, the third parameter indicates the size of the data in each task, and the fourth indicates where the *destination* data is stored. For `gather` and `reduce` the first and second parameters indicate the *destination* task rank and where the *destination* data is stored, the third parameter indicates the size of the data in each task, the last

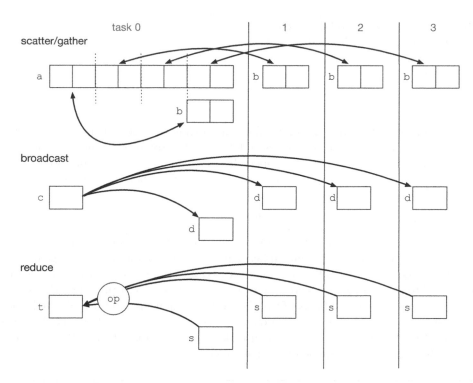

Figure 4.17: Four global communication patterns: scatter(0,a,2,b), gather(0,b,2,a), broadcast(0,c,size,d), reduce(0,s,size,op,t).

parameter indicates the *source* data, and for reduce the fourth parameter indicates the operation to be performed.

The following lines can be added to beginning of the pseudocode for the Game of Life to read the matrix and scatter it:

if $id = 0$ **then**
    read *grid* from disk
end
scatter(0, *grid*, $n * n/p$, $grid[1..n/p, 0..n-1]$)

The first parameter of the scatter indicates the source task, the second parameter points to the source data, the third parameter indicates the number of elements sent to each of p tasks (including the source), and the last indicates that the data is to be stored beginning in the second row of the grid array (skipping the ghost row). Note that *all* tasks make the same call to scatter, which results in each task (including the source) getting a block of n/p rows of matrix, in the case where $n \bmod p = 0$. MPI offers variations on collective communication functions that allow chunks to be sized differently for each task.

The following lines can be added to the end of the body of the outer **for** loop to gather and display results every d generations:

if *number of generations* mod $d = 0$ **then**
    gather$(0,\ grid[1..n/p, 0..n-1],\ n*n/p,\ dgrid)$
    if $id = 0$ **then**
        display $dgrid$
    end
end

The second parameter of the **gather** points to the data that is to be sent from a task, and the last parameter indicates where the gathered data is to be stored in the destination task. It would also be possible to allow task 0 to specialize in displaying results only, so that the time to display results could overlap with the calculations of further generations.

Global Communication: Broadcast and Reduction

The two other most common collective communication patterns, *broadcast* and *reduction*, can be illustrated with matrix-vector multiplication. First, let's look at a row-wise decomposition, and assume that the matrix is already distributed and the vector is replicated across all tasks. After each task completes its local matrix-vector multiplication, the resulting vector is distributed across all tasks. If we wish to have the result vector replicated, then the vector needs to be gathered and then broadcast, as shown in Algorithm 4.17. Vector c is gathered into task 0, then is broadcast to all tasks (including task 0). In MPI these two operations can be replaced by a single operation called an *all-gather*.

Algorithm 4.17: Matrix-vector multiplication with row-wise decomposition and message passing.

$nb \leftarrow n/p$ **//** **assume** n mod $p = 0$
for $i \leftarrow 0$ *to* $nb - 1$ **do**
    $c[i] \leftarrow 0$
    foreach *column j of a* **do**
        $c[i] \leftarrow c[i] + a[i,j] * b[j]$
    end
end
gather$(0,\ c,\ nb,\ c)$
broadcast$(0,\ c,\ n,\ c)$

The 2D decomposition is more complex to program, because it requires a reduction among tasks in each row followed by a broadcast to each column of tasks. This is an example that shows the need for collective communication across a subset of tasks. Algorithm 4.18 shows the implementation of the 2D decomposition, as illustrated in the task graphs of Figures 4.14 and 4.15. We assume for convenience a $n \times n$ matrix and the number of tasks is square $(p = q^2)$. This allows us to easily determine the coordinate of each block in the $q \times q$ grid of tasks, including those in the first column $(id = destID)$. The fourth parameter in **reduce** indicates the type of reduction operation, in this case a sum. We've added the last parameter in **reduce** and **broadcast** to indicate the group of tasks that participate in each collective communication. The reductions takes place among tasks in a row, with the result going to the first task in a row. These destination tasks then broadcast their piece of array c to tasks in the corresponding task column.

Algorithm 4.18: Matrix-vector multiplication with 2D decomposition and message passing.

```
// Assume that p = q², matrix is square, and n mod q = 0
q ← √p
nb ← n/q
for i ← 0 to n/q − 1 do
    c[i] ← 0
    for j ← 0 to n/q − 1 do
        c[i] ← c[i] + a[i, j] * b[j]
    end
end
rowID ← ⌊id/q⌋
destID ← rowID * q
group ← [destID..destID + q − 1]
reduce(destID, c, nb, sum, c, group)
group ← [rowID, rowID + q, .., rowID + (q − 1) * q]
broadcast(destID, c, nb, c, group)
```

Subset Sum

We'll conclude our look at message passing by implementing a distributed parallel subset sum, in Algorithm 4.19. As in the shared memory implementations we can agglomerate the tasks in each row, and as a result the matrix is partitioned into columns. Tasks can fill a row independently, but communication is required between rows. Recall that the first column of F always contains the value 1, and the remaining elements of each row are computed using values from the previous row. Therefore we partition the columns of F excluding the first column, and afterwards add the extra element to task 0 (lines 4–6). Tasks with columns $j \geq s[i]$ require elements from tasks that contain columns $j − s[i]$. Each task has columns $myfirst \leftarrow \lfloor id * S/p \rfloor + 1$ to $mylast \leftarrow \lfloor (id + 1) * S/p \rfloor$ (lines 2, 3), and depends on those columns $myfirst − s[i]$ to $mylast − s[i]$ that are greater or equal to 0 (lines 26, 28).

To begin, task 0 broadcasts the array s of numbers to all tasks. There are only two nonzero elements in the first row: at index 0 and $s[1]$. The former is set by task 0. Procedure findID in Algorithm 4.20 identifies the owner of a column for a generic block decomposition. Note that when findID is called the first parameter has to be adjusted to account for the fact that our partition leaves out the first element. The task that owns column $s[1]$ sets the corresponding element in row 0 to 1, using procedure findID to identify the owner (line 10). As iterations progress through the following rows tasks send portions of their row that are needed by other tasks, then they apply the sequential algorithm (solveRow, line 38 and Algorithm 4.20) to their elements. The reporting of results is left out of the algorithm. If only a yes/no answer is required, then task $p − 1$ can output the result, and matrix F can be reduced to only two rows with execution alternating between rows. If the subset is required then the distributed matrix F can be traced.

What is noticeable immediately from the algorithm is how much more code is required than in shared memory implementations. Each task makes use not only of local indices but also needs to refer to global indices such as $myFirst$ and $myLast$. The sender/receiver ranks need to be identified, along with the send and receive buffer locations and sizes. Each task has an array L to receive messages containing elements needed because of the $j − s[i]$ dependency. Further complicating matters, portions of a row may need to be sent to 1 or

---

Algorithm 4.19: Subset sum with message passing

Input: Array $s[1..n]$ of n positive integers, target sum S
Output: Completed array F
Data: Array $F[1..n, 0..nb-1]$ initialized to 0 (nb is number of columns owned by task), array $L[0..\lceil S/p \rceil - 1]$ to store received messages

1: broadcast($0,s,n,s$)
2: $myFirst \leftarrow \lfloor id * S/p \rfloor + 1$
3: $myLast \leftarrow \lfloor (id+1) * S/p \rfloor$
4: **if** $id = 0$ **then** //first block has one extra value
5:     $myFirst \leftarrow 0$
6:     $F[1,0] \leftarrow 1$
7: **end**
8: $nb \leftarrow myLast - myFirst + 1$
9: $nL \leftarrow \lceil S/p \rceil$
10: **if** $id = $ findID($s[1] - 1, p, S$) **then** $F[1, s[1] - myFirst] \leftarrow 1$

11: **for** $i \leftarrow 2$ *to* n **do**
12:     $id1 \leftarrow$ findID($myFirst + s[i] - 1, p, S$)
13:     $id2 \leftarrow$ findID($myLast + s[i] - 1, p, S$)
14:     **if** $id1 < p$ **then**
15:         **if** $id1 = id2$ **then**
16:             $myLocalBegin \leftarrow \max(0, \lfloor id1 * S/p \rfloor + 1 - s[i] - myFirst)$
17:             **send** $F[i - 1, myLocalBegin..nb - 1]$ to $id1$
18:         **else**
19:             $destBegin \leftarrow myFirst + s[i]$
20:             $destLast \leftarrow \lfloor (id1 + 1) * S/p \rfloor$
21:             $nb1 \leftarrow destLast - destBegin + 1$ // # of elements to send to $id1$
22:             **send** $F[i - 1, 0..nb1 - 1]$ to $id1$
23:             **if** $id2 < p$ **then send** $F[i - 1, nb1..nb - 1]$ to $id2$
24:         **end**
25:     **end**
26:     **if** $myLast - s[i] \geq 0$ **then**
27:         $id1 \leftarrow findID(myFirst - s[i] - 1, p, S)$
28:         **if** $myFirst - s[i] < 0$ **then**
29:             $myLocalBegin \leftarrow s[i] - myFirst$
30:         **else**
31:             $myLocalBegin \leftarrow 0$
32:         **end**
33:         **receive** from $id1$ into $L[myLocalBegin..nL - 1]$
34:         $nS \leftarrow$ size of message received
35:         $id2 \leftarrow$ findID($myLast - s[i] - 1, p, S$)
36:         **if** $id1 \neq id2$ **then receive** from $id2$ into $L[nS..nL - 1]$
37:     **end**
38:     solveRow($F, L, s, nb, myFirst, i$)
39: **end**

    // Trace F to return subset, or have task $p - 1$ print $F[n, myLast]$ to
    return yes/no

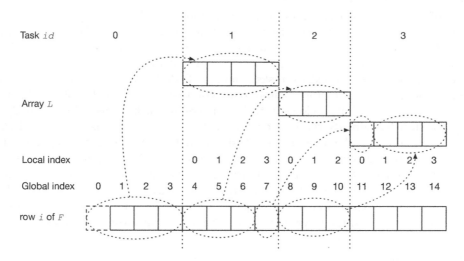

Figure 4.18: Subset sum with message passing for $S = 14$ and $s[i+1] = 4$.

2 other tasks (lines 15–24, 33–36), as seen in the example in Figure 4.18. Then there are boundary cases at the beginning and end of a row to be dealt with (lines 14, 16, 23, 26).

For example, in Figure 4.18 task 0 has elements with global indexes $myFirst = 0$ and

---

Algorithm 4.20: Procedures findID and solveRow for subset sum

---

```
// return rank of task that owns column j of array of length n in a
   decomposition into p blocks
Procedure findID(j, p, n)
   return ⌊(p * (j + 1) - 1)/n⌋
end

// Solve row i of F with array L for values needed from other tasks
Procedure solveRow(F, L, s, nb, myFirst, i)
   if id = 0 then
       F[i, 0] ← 1
       jstart ← 1
   else
       jstart ← 0
   end
   for j ← jstart to nb − 1 do
       F[i, j] ← F[i − 1, j]
       offset ← j − s[i]
       if offset ≥ 0 then //dependency in my array F
           F[i, j] ← F[i, j] ∨ F[i − 1, offset]
       else if offset + myFirst ≥ 0 then //dependency in array L
           F[i, j] ← F[i, j] ∨ L[j]
       end
   end
end
```

$myLast = 3$ of F, task 1 has elements 4 to 7, etc. Before tasks begin computing elements of row i they need to receive the data they depend on from row $i - 1$. For task 0, since $myFirst + s[i] = 4$ and $myLast + s[i] = 7$, which falls within the range owned by task 1, it sends $F[i-1, 0..3]$ to task 1 into $L[0..3]$. For task 1, $myFirst + s[i] = 8$ and $myLast + s[i] = 11$, which overlaps elements owned by tasks 2 and 3. Task 1 sends $F[i - 1, 0..2]$ to task 2 into $L[0..2]$ and $F[i - 1, 3]$ to task 3 into $L[0]$. For task 2, $myFirst + s[i] = 12$ and $myLast + s[i] = 14$, which both belong to task 3. Therefore task 2 sends $F[i - 1, 0..2]$ to task 3. Since task 3 has already received data from task 1, from the size $nS = 1$ of this message it knows to receive the message from task 2 in $L[1..3]$. Now that the dependencies from row $i - 1$ have been received, the elements of row i of F can be computed in `solveRow`. The horizontal dependence $F[i - 1, j]$ is within the array of each task. For tasks with $id > 0$ the diagonal dependence is either in the local array or in array L, whereas task 0 only has dependencies in its own array (and sets $F[i, 0] \leftarrow 1$). For our example with $s[i] = 4$, tasks have no dependencies within their own arrays since they have at most 4 elements.

4.8.3   Map-Reduce

The subset sum implementation using message passing illustrates the complexity of distributed data parallel programming. Even so, it is quite a simple application in that each task has the same fixed amount of work. It should perform well if it is executed on a computer that consists of identical compute nodes with all pairwise communication taking the same amount of time. If the computer is distributed in the cloud, however, then the processing power of nodes and the network performance may be uneven. The application would need to be modified to obtain optimal performance, making use of dynamic load balancing, possibly using the master-worker pattern. Fault tolerance may also need to be built in to deal with the likelihood of node failure in large scale distributed computers. These issues arising from cloud computing significantly complicate efficient implementation of parallel programs, even for simple algorithms. MapReduce deals with these issues for programs that make use of the embarrassingly parallel and reduction patterns.

Google's MapReduce framework was created to perform "simplified data processing on large clusters," as indicated by the title of the 2004 paper [17]. It has since been widely used in the form of the open source Hadoop framework in areas that range from the original purpose of indexing large collections of documents (such as the World Wide Web) to other applications in Big Data analytics [45]. Data processing workloads typically involve batch processing of very large collections of records. MapReduce applies a *map* function to a set of key-value pairs, where each pair represents the location of a value in the input, and the value itself. The map function is evaluated by map tasks that operate in parallel, each producing another set of key-value pairs. The next step, called a *shuffle*, gathers key-value pairs with the same key into key-list(value) pairs and sends each to a *reducer* task. Finally, the reducer produces the aggregated result for each key. Figure 4.19 shows a task graph for MapReduce.

The word count example that was presented in Chapter 1 is the prototypical MapReduce application. A map is applied to each line of text in one or more documents and the shuffle phase gathers resulting word-count pairs for each word, such as $\langle the, 4, 2, 1, 2, \ldots \rangle$ which represents counts of *the* for each line. The word-count reduce phase adds all the counts together for each key.

The shuffle phase is invisible to the programmer. The only programming needed is implementation of the map and reduce functions. The mapper for word count could look like:

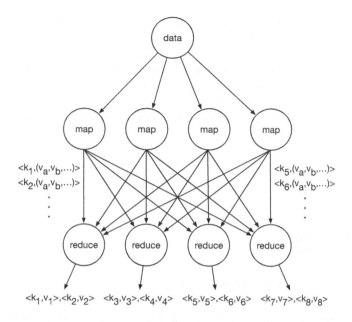

Figure 4.19: Task graph for MapReduce.

Procedure map(line)
    while line *has more words* **do**
        output(*word, 1*)
    end
end

For a large data set this would produce too many key-value pairs. This can be alleviated by inserting a *combine* phase after the map phase, which aggregates the values associated with each key from each map task. This phase combines all key-value pairs from each map with a user-defined function, which in this case would be the reduce function. This results in a word count for each word in the line processed by a map task. This lessens the workload of the shuffle step, which has fewer values to gather. The same reduce function is used to produce an overall word count:

Procedure reduce(*key, list(value)*)
    sum ← 0
    foreach *value in list* **do**
        sum ← *sum + value*
    end
    output(*key, sum*)
end

MapReduce frameworks take care of distributing data to mappers, shuffling data between map and reduce phases, and storing results, while ensuring load balance and fault-tolerance. It's not hard to see how good performance of the map phase could be achieved, by assigning workloads proportional to the computing capacity of processors and taking into account variations in communication times to different nodes. Nodes that were slow to compute and slow to communicate with could receive smaller workloads. With MapReduce we don't have a single parallel reduction, but instead have many small reductions (one for each key) that are computed in parallel. The largest overhead is the shuffle step, which can significantly impact the performance of MapReduce frameworks.

The k-means clustering algorithm is a good candidate for implementation using a MapReduce framework [79]. In each iteration each mapper takes a vector and the array of cluster centers, and finds the closest center, producing a key-value pair (*centerIndex*, *vector*). The reducer computes the sum of vectors assigned to each center, and outputs the index of the center as a key, and the pair (sum, number of vectors) as a value.

> **Procedure** map(*vector, cluster*)
>   find closest cluster center to *vector*
>   output(*centerIndex, vector*)
> **end**

> **Procedure** reduce(*centerIndex, list(vector)*)
>   *sum* ← 0
>   **foreach** *vector in list* **do**
>   *sum* ← *sum* + *vector*
>   **end**
>   output(*centerIndex, sum,* `sizeof`*(list)*)
> **end**

The result of the reducers can be used to calculate the mean, which is then stored in the *cluster* array to be used in the next iteration. Recall that the framework's shuffle phase takes care of gathering all vectors assigned to the same cluster center. With one mapper per vector this would produce a huge communication burden on the shuffle. A better approach would be to agglomerate the mapper tasks, where each new mapper operates on a group of vectors:

> **Procedure** map(*list(vector), cluster*)
>   **foreach** *vector in list* **do**
>   find closest cluster center to *vector*
>   output(*centerIndex, vector*)
>   **end**
> **end**

The framework would then use the **reduce** function to combine the results for each mapper. This is the reason the **reduce** function outputs the sum and number of vectors separately, instead of the mean, in order to allow partial sums to be computed for each mapper.

Guidelines for Distributed Memory Programming

We've only had a taste of distributed memory programming in this section. It adds to the complexity of SPMD programming the need to manage communication between tasks. Frameworks such as MapReduce hide this complexity for certain application patterns. We can highlight two important guidelines:

1. *Data distribution should minimize communication*

2. *Global communication routines should be used when appropriate*

As we saw above for matrix-vector multiplication, there are choices to be made regarding the distribution and replication of data. These choices should be guided by the goal of minimizing communication. We'll revisit this issue in Chapter 5. MPI programmers can be tempted to use send/receive operations in cases that would be better handled by using a global communication routine, as discussed above. This issue is nicely discussed in the paper *Send-receive considered harmful* [29].

4.9 CONCLUSION

We've explored in depth a number of commonly used program structures relevant to the three machine models presented in Chapter 2. It would be difficult to assemble a complete catalog of program structures. The goal of this chapter and the previous one has been to provide a guide to creating a parallel decomposition and implementing it for a particular machine model. Given a particular machine there are usually several program structures that can be considered. With this chapter as a guide, one or more suitable structures can be chosen, even before the programming language is considered. Given the hierarchical nature of parallel computers, as we saw in Chapter 2, it is worth considering combining several program structures (see Exercise 4.26). The SIMD program model should always be considered, as most processors support SIMD execution in some fashion. If the problem is large enough to target a cluster, then a coarse decomposition using a straightforward message passing implementation, can be combined with a fine grained decomposition for each compute node, using a suitable shared memory program structure. This hybrid approach has been successful using MPI and OpenMP [20].

4.10 FURTHER READING

Skillicorn and Talia give a helpful classification of parallel programming models in *Models and Languages for Parallel Computation* [66]. A more recent paper by Diaz et al. surveys the parallel programming models for multicore and manycore processors [20]. Chapters 5 and 6 of Mattson et al.'s *Patterns for Parallel Programming* [50] discuss parallel program design patterns. An early use of the fork-join structure was in the Cilk programming language [25], and Intel provides good online tutorials for the current Cilk Plus language. Java's fork-join framework is clearly presented by its author Doug Lea [46], and practical aspects of its use are documented in *Fork-Join Parallelism in the Wild* [16]. While parallel loops can be used in Cilk Plus, they have been most successful in the OpenMP API. The `openmp.org` website is a good place to start, with links to books and online resources. This is also a good place to learn about how OpenMP supports tasks with dependencies. A related programming model that supports tasks with dependencies is the Barcelona Supercomputing Center's OmpSs [22]. Herlihy and Shavit's *The Art of Multiprocessor Programming* [35] provides a throrough and very readable study of the theory and practice of concurrency for shared memory programming. There are many resources on programming manycore processors, but an important starting point is understanding the programming model, which Garland and Kirk's *Understanding Throughput Oriented Architectures* [28] explains. Gropp et al.'s book *Using MPI* [31] is an excellent resource for distributed memory programming with MPI. Finally, there is an abundance of online resources on MapReduce and related frameworks. Stewart and Singer present an interesting comparison of two parallel program structures in *Comparing Fork/Join and MapReduce* [71].

4.11 EXERCISES

There are several options for implementing the programming models presented in this chapter. SIMD is supported by extensions to several models, including Cilk Plus and OpenMP, and in the CUDA SIMT model. Note that not all SIMD examples may be able to be vectorized by the compiler, when using Cilk Plus or OpenMP. Cilk Plus and Java both have good support for fork-join programming. Cilk Plus and OpenMP both support loop parallelism. OpenMP, OmpSs, and Intel's Thread Building Blocks support tasks with dependencies. SPMD programming can give good results with OpenMP and/or MPI. Hadoop is a popular open source framework for MapReduce programming.

For all implementations, generate or find suitable input data, and make sure to vary the problem size. A small input size is good for debugging, but larger problem sizes are needed to see any performance gain. See the discussion in Chapter 5.

4.1 Choose two parallel programming models in practical use today and contrast their support for parallel program structures. Are there structures supported that aren't included in this chapter?

4.2 Compare the performance of sequential matrix-vector multiplication in two versions: one with iteration over rows in the outer loop and the other iterating first over columns. Explain the performance difference observed. Now implement SIMD row-wise and column-wise matrix-vector multiplication and compare their performance. How is the performance difference between SIMD versions related to the difference between sequential versions?

4.3 Implement SIMD subset sum. Time each outer loop iteration for sequential and parallel execution. How does the relative performance of the two vary among iterations? Experimentally study the effect of the magnitude of S on overall performance of SIMD execution.

4.4 Draw the array access pattern for Blelloch's scan algorithm (Section 3.4), as was done for Hillis and Steele's algorithm in Figure 3.11. Then express Blelloch's algorithm using the SIMD notation of this chapter.

4.5 Express Hillis and Steele's scan algorithm(Section 3.4) using the SIMD notation of this chapter. Compare both scan algorithms using the guidelines of Section 4.2

4.6 Implement both SIMD scan algorithms (Exercise 3.4). How does their performance compare?

4.7 Parallelize fractal evaluation (Algorithm 4.7) using a fork-join framework. This will require recasting the outer loop as recursive calls. How does varying the sequential cutoff affect performance? Look at the result of increasing the thread pool beyond the number of physical and virtual (with hardware multithreading) cores.

4.8 Implement the fork-join merge sort of Algorithm 4.6 in two stages. First, use a sequential merge instead of `parMerge`. Then implement the parallel merge, and experimentally determine the performance improvement. How does varying the sequential cutoffs in `parmergeSort` and `parMerge` affect performance?

4.9 Modify the fork-join merge sort implementation of Exercise 4.8 to use a temporary array to perform each merge, instead of merging alternatively between two arrays. How does this change affect performance?

4.10 Parallelize fractal evaluation (Algorithm 4.7) using a parallel loop programming model. Experiment with different loop schedules. How does the performance compare to the fork-join implementation of Exercise 4.7?

4.11 Consider the second decomposition of the histogram evaluation in Figure 3.1b. Coarsen the decomposition into fewer tasks and implement it using a parallel loop programming model in two different ways:

    a. Using a private array for each thread and summing them at the end.

    b. Using a shared array and a critical section to protect shared access.

    Compare the performance of these two versions.

4.12 Implement subset sum (Algorithm 4.11) with a programming model that supports tasks with dependencies. How does its performance compare with a parallel loop implementation?

4.13 Parallelize fractal evaluation using SPMD programming in three ways: a) using static load balancing with contiguous blocks of rows; b) using cyclic static load balancing (Algorithm 4.12); c) using master-worker load balancing (Algorithm 4.15). Compare the performance of the three approaches.

4.14 Modify the SPMD merge sort Algorithm 4.13 so that it can handle an arbitrary number of threads, not just a power of two. This means that the number of threads assigned to each parallel merge may not be the same.

4.15 Implement three versions of SPMD merge sort: a) Algorithm 4.13; b) modified to allow any number of threads (Exercise 4.14); c) further modified to relax the constraint on n, so that $n \bmod nt \neq 0$ is allowed.

4.16 The Boolean Satisfiability Problem (SAT) is a fundamental problem in computational complexity. It was the first known NP-complete problem, and has applications in circuit design and automatic theorem proving. A SAT problem is a conjunction of clauses $C_1 \wedge C_2 \wedge C3 \wedge \ldots$, where each clause is a disjunction of Boolean variables, and these variables may be negated. An example is $(x_1 \vee x_2 \vee \neg x_4) \wedge (x_2 \vee \neg x_3) \wedge (x_3 \vee \neg x_1 \vee x_4)$. This example is satisfiable, where $x_1 = x_2 = x_3 = x_4 = T$ is one solution. Sophisticated heuristic algorithms based on tree search have been developed, but the simplest solution is to try all possible 2^n assignments of values to n variables. Design and implement two such brute force algorithms, using either shared or distributed memory models, to count the number of assignments that satisfy the formula. In the first version, assign an equal number of assignments to each task using the SPMD model. In the second version, balance the load using the Master Worker model. There are many online SAT resources. Find or generate problem instances that are small enough to be solvable in a reasonable time sequentially.

4.17 Implement and compare the performance of three versions of reduction using CUDA: Algorithm 4.14, and two other versions discussed in Section 4.2.

4.18 Implement a master-worker parallelization of fractal evaluation using MPI. Find a chunk size that trades off load balancing with minimization of communication.

4.19 Implement and compare the performance of 1D and 2D decomposition of matrix vector multiplication using MPI (Algorithms 4.17 and 4.18), as a function of matrix size.

4.20 Implement the Game of Life using MPI (Algorithm 4.16 with improvements for I/O). How does the performance vary with the problem size?

4.21 Implement Algorithm 4.19 for the subset sum using MPI. Does the use of nonblocking vs. blocking "send" affect performance?

4.22 Implement the parallel k-means decomposition of Figure 3.8 using MPI. You'll need to figure out how to distribute documents to tasks. You'll also need to determine convergence and communicate this information to all tasks.

4.23 Implement the k-means algorithm using a MapReduce framework. If the results of Exercise 4.22 are available for the same parallel computer, how does their performance compare?

4.24 The Naive Bayes classifier is a simple but powerful tool in machine learning. Given a data set where each point has k attributes $X = \{x_1, x_2, ... x_k\}$ and a label C, we can calculate the probability of attribute x_i appearing in the data set, $P(x_i)$, and also the conditional probability of observing attribute x_i given label C, $P(x_i \mid C)$. This allows us to predict the label of a new unlabeled point by using Baye's theorem to determine the probability of observing label C given a value of x_i: $P(C \mid x_i) = P(x_i \mid C)P(x_i)$.

Use a MapReduce framework to train a Naive Bayes classifier, that is compute $P(x_i)$ and $P(x_i \mid C)$, using a publicly available data set, such as the UCI machine learning repository.

4.25 A parallel decomposition of the Floyd-Warshall all-pairs shortest path algorithm was the subject of Exercise 3.9. Implement this algorithm three times using three different parallel program structures, one of which must be for distributed memory computers. Compare the three versions, discussing how they differ in terms of solution time and execution time.

4.26 It's not uncommon to find two or more program structures being used together to implement an algorithm, as parallel computers with more than one compute node feature all three machine models we've discussed. Choose an algorithm discussed in this chapter (including the exercises) and implement it with a hybrid of two machine models, such as parallel loops and SIMD, for example.

CHAPTER 5

Performance Analysis and Optimization

The vast space of possible solutions makes parallel computing both challenging and enjoyable. We have already seen that a good parallel decomposition results in a large number of independent tasks that grows with the problem size. The solution is further developed using a parallel program structure that is suited to the targeted computing platform. Analysis of the expected performance of a parallel algorithm is crucial, since the goal of parallel computing is increased performance.

5.1 WORK-DEPTH ANALYSIS

A good first step in performance analysis is to evaluate a parallel decomposition by examining the *work* and *depth* of its task graph. Let's consider the first level in the histogram task graphs from Chapter 3 in isolation. This involves calculating the complexity of each message, which requires parsing all characters followed by a sum over the alphabet to determine the number of distinct characters in the message. If this is done sequentially for n messages, then it will take time $\sum_{i=1}^{n} c_1 L_i + c_2$ (in $O(nm)$), where L_i is the length of message i, m is the length of the longest message and c_1, c_2 are constants. When these tasks are executed in parallel the runtime is given by that of the task that completes last (see Figure 4.1b), and will take time $\max_{i=1}^{n} c_1 L_i + c_2$ (in $O(m)$). We can conclude that parallel execution of this step is $O(n)$ faster than sequential execution.

Consider the first histogram decomposition (Figure 3.1a). Sequential execution of all tasks would take $O(nm + 16n + 16) = O(nm)$. For parallel execution the tasks in the second level depend on completion of all tasks in the first level, and all tasks in each of the first two levels would execute in parallel, resulting in a time of $O(m+n)$, where we neglect the gather step since it takes constant time with respect to n and m. Therefore parallel execution is $O(nm/(m + n))$ faster than sequential execution.

The sequential execution of all tasks in the second histogram decomposition (Figure 3.1b) would take $O(nm + n + n) = O(nm)$. For parallel execution we'll first treat the merge task as an $O(n)$ sequential reduction of the histograms produced by the tasks in level 2. Unlike in the previous decomposition, here execution of tasks in the first two levels can take place in parallel. Since some tasks in level 1 will finish earlier than others, their child tasks in level 2 will start executing before all tasks in level 1 complete. There are n possible paths of execution through the first two levels. The execution time is given by the maximum execution time of all the paths. The parallel execution time is $O(m + n)$, which is the same as for the first histogram. We can do better by performing the reduction in parallel. First, let's define some terms.

Definition 5.1 (Depth). *The parallel execution time is given by the **depth** of the task graph. To compute the depth we need to consider all possible paths in the task graph, and the accumulated times of the tasks in each path; the depth is given by the longest time of all the paths. If all the tasks take the same time then the depth is the time of the longest path, also called the **critical path**. No matter how many computational resources are available to execute the tasks in the graph, the solution cannot be computed in less time than the depth.*

Definition 5.2 (Work). *The sequential execution time of a task graph is called the **work**. It is not necessarily the same as the execution time of an efficient sequential algorithm. As we shall see, the work of parallel decompositions is often greater than that of the corresponding sequential algorithm. A parallel algorithm is **work-efficient** if its work has the same complexity as the best sequential algorithm.*

Definition 5.3 (Parallelism). *The upper bound on the performance gain of a parallel algorithm can be determined by its **parallelism**, which is given by the work divided by the depth.*

Consider the reduction task graph in Figure 3.5. Each task takes constant time, so the work is $O(n)$. Since this is the same complexity as sequential reduction the parallel reduction is work efficient. The depth is clearly $O(\log n)$, and the parallelism is then $O(n/\log n)$. Applying parallel reduction to the final step in the second histogram decomposition results in $O(m + \log n)$ depth, with the work unchanged at $O(nm)$. This gives higher parallelism than the first histogram decomposition. The difference is more evident if m is restricted to be a small value, such as the 140 character limit of Twitter. Then the first decomposition has only $O(1)$ parallelism, whereas the second has $O(n/\log n)$ parallelism.

Merge Sort

If we apply work-depth analysis to merge sort we can formally express the inefficiency of the naive divide and conquer algorithm discussed in Section 4.3. Recall that the recursive calls spawn new tasks, but the merges take place sequentially. The work of the naive algorithm is identical to sequential merge sort, $O(n \log n)$. The depth is given by the recurrence relation $d(n) = d(n/2) + O(n)$, where the second term comes from the merge. Solving this recurrence gives a depth of $O(n)$, which means the parallelism is only $O(\log n)$. This clearly shows the bottleneck of the $O(n)$ merge. It can be shown that the parallel merge of Algorithm 4.6 has $O(n)$ work and $O(\log^2 n)$ depth [13]. Applying parallel merge to merge sort leaves the work unchanged at $O(n \log n)$, and leads to the recurrence relation for the depth of $d(n) = d(n/2) + O(\log^2 n)$, which yields $O(\log^3 n)$ depth. The parallelism is then $O(n/\log^2 n)$, which is much better than the $O(\log n)$ of the naive merge.

Comparing Two Scans

The scan operation provides an interesting case of two competing parallel algorithms. The first, from Hillis and Steele, (Figure 3.11) is a SIMD algorithm:

```
for k ← 0 to log n − 1 do
    j ← 2^k
    {a[i] ← a[i − j] + a[i] : i ∈ [0..n) | i ≥ j}
end
```

This algorithm performs an inclusive scan in place. An exclusive scan can easily be obtained by using 0 followed by the first $n - 1$ values from the inclusive scan. The task graph has $n - 2^k$ independent tasks (each doing an addition) at each level, and tasks in one level complete before the next level starts. The depth is $O(\log n)$, and the work is $O(n \log n)$, so it is not work-efficient. The parallelism is $O(n)$.

Blelloch's tree-based scan has two traversals of the reduction tree, so it has the same asymptotic complexity as reduction, that is $O(n)$ work, $O(\log n)$ depth, and hence $O(n/\log n)$ parallelism.

Which algorithm is better? Hillis and Steele's has more parallelism, as there are more tasks at each level than in the second algorithm. It isn't work-efficient however, as it does much more work than a sequential scan, whereas Blelloch's is work-efficient. To better judge these algorithms we need to consider the computing platform, in particular the number of cores.

5.2 PERFORMANCE ANALYSIS

The work-depth analysis of the task graph of a parallel decomposition provides a rigorous analysis of the potential parallelism. It's worth emphasizing again the virtue of leaving out consideration of the computing platform in the first stage of algorithm design. The next stage is to analyze the performance of the algorithm once tasks are agglomerated and mapped to a computing platform. As we've discussed, agglomeration coarsens the decomposition, which reduces the overhead involved with mapping tasks to processors, and can also reduce communication overhead in a distributed implementation. An important choice that needs to be made is to identify the number of cores to use.

An upper bound on the number of cores is given by the maximum *degree of concurrency* of the task graph. This is the maximum number of tasks that can be executed concurrently at any time during execution. If the number of cores were increased to above the maximum degree of concurrency then there would always be idle cores and execution time would not be improved. For example in shared memory reduction, the degree of concurrency is $n/2$, which is the number of leaf tasks.

5.2.1 Performance Metrics

Whether the mapping of tasks to cores is done by the programmer or by system software, the performance of an algorithm can be studied for a given number of cores of a computing platform. There are a number of performance metrics that have been devised, but we will focus on the most important and commonly used: speedup, cost, efficiency, and throughput metrics.

Speedup

There are two definitions of speedup, and it is important to identify which one is being used.

Definition 5.4 (Relative Speedup). *The relative speedup is given by the ratio of the execution time of the parallel program on one core to the time on p cores.*

The parallelism of work-depth analysis gives an upper bound on the relative speedup since it assumes an unlimited number of cores are available. The relative speedup a useful metric to express the scalability of the algorithm.

Definition 5.5 (Absolute Speedup). *The absolute speedup is given by the ratio of the execution time of the best known sequential algorithm to the execution time of the parallel algorithm on p cores.*

The second definition can be used to honestly advertise the performance of your parallel algorithm. In many algorithms the work done by the parallel algorithm is greater than that of the best sequential algorithm (recall that if it's greater by more than a constant factor then it is work-inefficient). This means that the parallel algorithm executed on a single core will typically take longer than a proper sequential algorithm, and thus the relative speedup will be higher than the absolute speedup. One should therefore be very clear about which definition of speedup is being used when documenting the performance of an algorithm.

If there is W amount of work to do, and p cores are available, then the best we can do is assign W/p units of work to each processor, resulting in a speedup of p (see Section 4.1). There are two exceptions to this upper bound on speedup. In exploratory algorithms where the amount of work can be much less than the sequential algorithm, so-called super-linear speedup greater than p is possible. This is the case with parallel tree search, where the tree has been partitioned and assigned to cores. The amount of work required to find a solution can be much less than the sequential search, therefore the speedup can exceed p. The other exception arises from data parallel programs where the portion assigned to each core fits into high speed cache memory, which leads to super-linear speedup when compared to the sequential program operating on the entire data that does not fit in the cache.

In practice, sadly, most parallel programs have sub-linear speedup, which can be due to a number of factors. One of them is due to an algorithm that is not *cost efficient*.

Cost

Computing resources are not free, and increasingly are rented by the CPU-hour from cloud service providers. If the parallel program consumes significantly more CPU-hours than a sequential program then the desired performance may only be achievable at an unacceptable cost.

Definition 5.6 (Cost). *The cost of a parallel algorithm is the product of the asymptotic execution time and the number of cores.*

The cost is related to the work of the work-depth model but they are not identical. Work does not consider cores at all, so cannot account for the fact that once tasks are mapped to cores, these cores may be idle some of the time. Consider parallel reduction, where there is $O(n)$ work. If $p = n/2$ cores are used in a shared memory implementation then they are all busy only in the first step. In each subsequent step half the cores become idle. The cost of $O(p \log p)$ takes into consideration that p cores were available for $\log p$ steps, even if many of them were idle much of the time. As with work, one can define a *cost efficient* algorithm as one whose cost has the same complexity as the best sequential algorithm. Clearly, reduction with $n/2$ cores is not cost efficient, as the cost is $O(\log n)$ times greater than the sequential algorithm.

A cost inefficient algorithm doesn't only lead to an expensive implementation, but also suggests that a better implementation may be possible. Shared memory reduction with $p = n/2$ cores gives a speedup of $O(n/\log n) = O(p/\log p)$. Instead of choosing the number of cores equal to the degree of concurrency, we can instead use fewer cores. Each core begins by summing n/p numbers, then the same tree-based reduction is used to sum the resulting p numbers. The time of this alternative algorithm is $O(n/p + \log p)$, so its cost is

$O(n+p\log p)$. If n is greater than $p\log p$ then this algorithm is cost efficient and its speedup is $O(n/(n/p)) = O(p)$, which is optimal.

Efficiency

Definition 5.7 (Efficiency). *Efficiency is defined as the ratio of the speedup to the number of cores.*

An algorithm with linear speedup (p) has an efficiency of 1. In practice the efficiency of parallel programs is usually less than 1. Efficiency is also given by the ratio of the execution time of the sequential algorithm to the cost. This clearly shows the importance of a cost efficient algorithm, which results in an efficiency that is within a constant factor of 1 for any number of cores.

Throughput Metrics

One important limitation of the speedup metric, when applied to experimental results, is that both sequential and parallel programs must be executed on identical cores. This may not be meaningful, as in the case of GPU programs. Sequential programs are not run on GPUs and the meaningful comparison is between sequential execution on a CPU core and parallel execution on GPU cores. The problem with this type of speedup is that it depends on the type of CPU and GPU cores employed, and as a result it is difficult to compare speedups from experiments on different CPUs and GPUs. Speedup may also not be possible to measure, for instance if the entire data set does not fit into the memory of a single compute node which prevents meaningful sequential timing from being obtained.

An alternative to speedup is a throughput metric, such as FLOPS (FLoating point Operations per Second) for numerical applications and CUPS (cell updates per second) for dynamic programming applications, which addresses these limitations of speedup.

Let's take two hypothetical scenarios where the subset sum algorithm is parallelized for a GPU using a SPMD algorithm. Both scenarios use the same implementations and the same problem size. In the first case, the sequential implementation takes 30 minutes to run on CPU1 and the parallel implementation takes 1.5 minutes to run on GPU1. Additionally, a loop parallel implementation is run on CPU1 using 4 cores, and takes 12 minutes. What is the speedup? In the literature you will find speedups for GPUs based on sequential and multithreaded CPU execution. This would give speedups of $30/1.5 = 20$ and $12/1.5 = 8$, respectively.

Now consider the second case, where sequential execution on CPU2 takes 40 minutes, parallel execution on CPU2 takes 15 minutes, and parallel execution on GPU2 takes 2 minutes. This gives speedups of 20 and 7.5, based on sequential and parallel CPU execution respectively. Comparing the first set of speedups would suggest that the same speedup of 20 was obtained on both GPU1 and GPU2. Comparing the second set of speedups suggests that GPU2 produces slightly lower performance (7.5 vs. 8). In reality, of course, the performance difference is greater. Let's say the GPU1 run results in 20 Giga CUPS (GCUPS). This means that the run on GPU2 gets $(1.5/2) \times 20 = 15$ GCUPS, since the workload is the same. Publishing the GCUPS figures gives the correct relative performance on the two GPUs.

A warning about throughput metrics is in order, as they can be abused. Recall in our discussion above of the work metric that parallel algorithms often perform more computation to solve a given problem than the best sequential algorithm. Therefore, by itself, a higher FLOPS rate doesn't mean better performance, if it is based on an operation count of the parallel program. Instead, the operation count should be based on the best known sequential algorithm.

Scan Algorithms Redux

Let's re-examine the competing scan algorithms. If we can use an n-way SIMD processor for the Hillis and Steele algorithm then the parallel runtime is $\log n$ and the speedup is $n/\log n$. If we use $p = n/2$ cores for Blelloch's scan then the speedup is $n/(2\log n)$. These are both cost inefficient, at $O(n\log n)$.

What if fewer cores than the degree of concurrency are used? For the Hillis and Steele algorithm if n is greater than the SIMD width p, then the sums at each level will be done one chunk of p elements at a time. The speedup becomes $O(n/((n/p)\log n)) = O(p/\log n)$, and the cost, $O(n\log n)$, is inefficient. The performance of the Blelloch algorithm for $p < n/2$ depends on how tasks are scheduled to cores. A simple schedule proceeds level by level. When there are more concurrent tasks than cores, tasks assigned to each core are executed sequentially. At the first level each core executes $n/(2p)$ tasks sequentially, then $n/(4p)$ at the next level, and so on until the number of tasks reaches p, when the algorithm proceeds with each task executing in parallel at each level. This results in $O((n/p)\log n)$ time, which is the same complexity as the competing algorithm.

A better approach is to follow the modification to the reduction algorithm described above. We'll use the shared memory SPMD model. As with reduction, we can assign the n numbers to $p \ll n$ cores, and insert an initial step where each core performs a scan of its n/p numbers. The sum of each partial array produces p values. A parallel scan then proceeds with these p values, then in a final step, each core (except the one with thread $id = 0$) adds an offset from the scan to its subarray. The parallel scan in Algorithm 5.1 could use either of the two algorithms we've seen.

Algorithm 5.1: SPMD scan

```
// assumes n divisible by p
shared a, b
start ← id * n/p
sum ← 0
for j ← 0 to n/p − 1 do
    sum ← sum + a[start + j]
    a[start + j] ← sum
end
b[id] ← a[n/p − 1] // sum of all values in sub-array
scan(b)
if id > 0 then
    for j ← 0 to n/p − 1 do
        a[start + j] ← a[start + j] + b[id − 1]
    end
end
```

The parallel execution time of the two **for** loops in Algorithm 5.1 is $2\lceil n/p \rceil$, counting loop iterations. If we use the Hillis and Steel scan the total execution time is $2\lceil n/p \rceil + \log p$, and the competing algorithm has an execution time of $2\lceil n/p \rceil + 2\log p$. The latter algorithm has an extra $\log p$ steps but both have identical asymptotic execution time. They both have a cost of $O(n + p\log p)$, so they are cost efficient if n is greater than $p\log p$. In practice, the

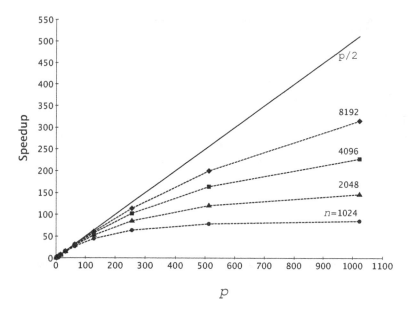

Figure 5.1: Speedup as a function of number of cores for different problem sizes, for Hillis and Steele SPMD scan. Also shown is $p/2$ upper bound.

choice of one algorithm over the other will depend on how implementations of each can be optimized on a given platform, taking into account factors such as cache locality.

Strong and Weak Scaling

It's instructive to look at the speedup curve of the scan algorithms as a function of problem size. Since the algorithms are cost efficient, the speedup will be within a constant factor of optimal linear speedup. The speedup of the SPMD Hillis and Steele algorithm is $n/(2\lceil n/p\rceil + \log p)$, which approaches $p/2$ as n increases to a value much larger than p. This asymptotic speedup is less than p because of the need for two traversals of the array in the first and third steps. For n closer to p, the $\log p$ term from the scan in the second step reduces the speedup even further.

The speedup plot for this algorithm in Figure 5.1 shows a common trend in the speedup of parallel programs, where increasing the problem size increases the speedup. For our example this occurs because the $\log p$ term makes a smaller contribution to the execution time as n increases. One of the common mistakes made by beginning parallel programmers is to evaluate the performance of their program using a problem size that is too small. In another view of the same results, Figure 5.2 shows how the efficiency drops with increasing p.

Each curve in Figures 5.1 and 5.2 is for a fixed problem size. This reflects the case where more cores are being used to reduce the execution time for a fixed size. In contrast, more cores are often used to solve larger problems in roughly the same time. Figure 5.3 shows that if we scale the problem size as $n = 512p$ the execution time increases by less than 1% as p increases to 1024. The corresponding plot of efficiency in Figure 5.4 shows that if we keep n/p fixed as we increase p we can keep the efficiency close to the upper bound of $1/2$, as it only drops by a factor of $\log p$. This plot is an example of *weak scaling*, where the amount of work per core is kept fixed as the number of cores increases. Figures 5.1 and 5.2

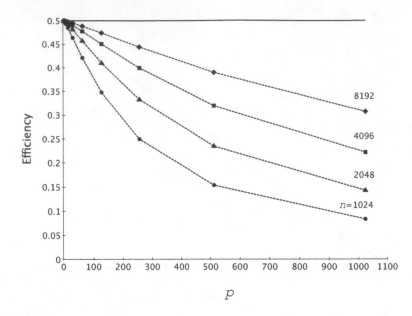

Figure 5.2: Efficiency as a function of problem size for a Hillis and Steele SPMD scan. Also shown is 1/2 upper bound.

show *strong scaling*, where the overall problem size is kept constant as the number of cores increases.

5.2.2 Communication Analysis

There are numerous factors that can reduce speedup. For reduction and scan, it is the need for $\log p$ steps. In general there are many barriers to performance. A significant barrier for parallel algorithms for distributed memory computers is communication overhead.

The time required to send a message over a communication link is usually modeled as $t_{comm} = \lambda + m/\beta$. The first term is for the latency λ, which includes the hardware latency and software overhead to setup the communication. The second term is the time required to transmit a message of length m over a link of bandwidth β. At the limit of large or small messages the first or last terms, respectively, are sometimes dropped to simplify the analysis.

The communication pattern of a distributed parallel algorithm forms what can be called a *virtual topology*. For example, a reduction forms a tree topology, and grid based applications such as Conway's Game of Life form a nearest-neighbor mesh topology. The actual communication time will depend on the embedding of the virtual topology onto the *physical topology* of the parallel computer. We will only consider the virtual topology, assuming no loss of performance from embedding on a network. In practice, communication patterns can be designed for the characteristics of the computing platform, taking factors such as the physical topology and also that some cores will be "closer" together than others. For example, in a cluster of multicore nodes, cores can communicate more rapidly with neighbors in the same node than with cores in other nodes. However, too much intra-node communication can overwhelm the capacity of the memory bus. Such considerations are beyond the scope of this book.

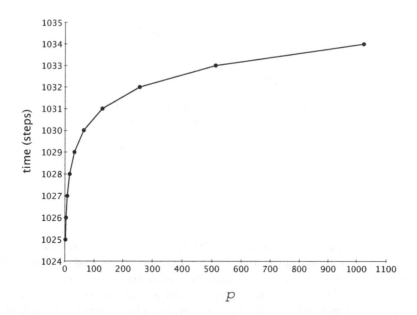

Figure 5.3: Weak scaling execution time for a Hillis and Steele SPMD scan. Problem size is scaled as $n = 512p$.

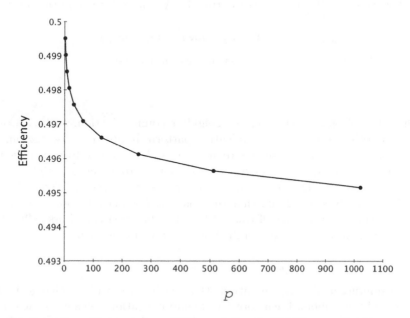

Figure 5.4: Weak scaling efficiency for a Hillis and Steele SPMD scan. Problem size is scaled as $n = 512p$.

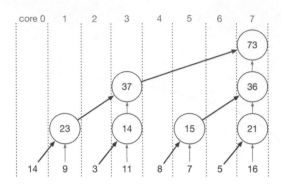

Figure 5.5: Reduction tree mapped to 8 cores.

Reduction

Let's consider the ubiquitous reduction pattern, for the case where each core has a single value and the sum of all values will be stored in a single core. The tasks in the reduction tree can be mapped to cores as shown in Figure 5.5. Half the cores send their values to the other cores, where the first level tasks execute. Then $p/4$ of the cores perform the next level tasks after receiving values from the other $p/4$ cores. This continues, with half of the remaining cores dropping out at each level, until one core has the final result. Compare with the task graph of Figure 3.5a, and you'll see that at each level half the dependencies are between tasks mapped to the same core (gray arrows), whereas the other half are between cores (bold arrows) and so require communication. At each stage $p/(2i)$ communications take place in parallel, so the communication time is the same as that of sending a single message. Therefore the overall communication time is $(\lambda + m/\beta) \log p$, and the total time is

$$t_{\text{reduction}} = (\sigma + \lambda + m/\beta) \log p = O(m \log p),$$

where σ is the time to add to values and m is the size of a value.

Game of Life

Grid-based parallel applications with nearest-neighbor communication, such as Conway's Game of Life, use another common communication pattern. In our row-wise agglomeration in Algorithm 4.16, each block is assigned to a core, and the cores form a ring virtual topology. Each core sends its upper and lower boundary rows to its left and right neighbors and receives its two ghost rows from the same neighbors, as can be seen in Figure 4.16. If we assume that links are bi-directional and that there is only one network port per core, then the communication time for exchange of rows of length n between neighbors is $2(\lambda + n/\beta)$, since each core sends 2 messages. The overall time per iteration is then:

$$t_{1D} = O(n^2/p + n).$$

It's worth considering a 2D agglomeration. Here each core would exchange boundary information with eight neighbors. Each core has 4 communication steps with each message size proportional to $n/\sqrt{p}$, neglecting the communication of four single values for the four corners. This gives a time per iteration of:

$$t_{2D} = O(n^2/p + n/\sqrt{p}).$$

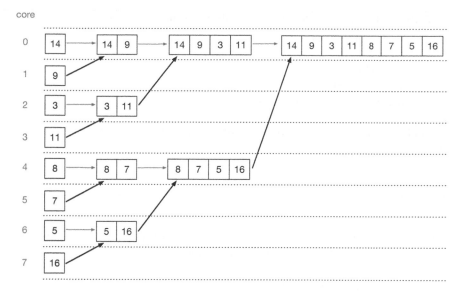

Figure 5.6: Gather tree mapped to 8 cores.

Both versions are cost efficient but the 2D decomposition has a communication overhead of lower complexity.

Note also that the 2D decomposition has better weak scaling. To keep the work constant as the number of processors increases, n increases as a function of $\sqrt{p}$. For the 2D decomposition the communication stays constant as p increases, so the overall execution time is unchanged as a function of p. This means that the efficiency is of $O(1)$, and we can say that this algorithm is *perfectly scalable*. In contrast, the 1D decomposition communication increases linearly with n, so the weak scaling efficiency drops by a factor of $\sqrt{p}$ as p increases.

Matrix-Vector Multiplication

Matrix-Vector multiplication provides another opportunity to compare one and two dimensional agglomeration and to estimate the cost of gather and broadcast operations. In the 1D agglomeration the matrix is distributed row-wise to cores and the vector is replicated. Communication is required to gather the partial result vectors from all cores and then to broadcast the result vector to all cores. Both of these collective operations can be performed in a similar manner to a reduction. The broadcast of an array of length n can use the same communication pattern as a reduction, except in the reverse direction, and without the sum operation, which results in a time of:

$$t_{\text{broadcast}} = (\lambda + n/\beta) \log p = O(n \log p).$$

The gather behaves a little differently in that the message size increases each step, and that the order of the cores assigned to the row-wise agglomeration must be followed, as seen in Figure 5.6. To simplify the analysis we'll assume that p is a power of 2 and that n is divisible by p. In the first step cores with an odd *id* send their sub-arrays to cores with even *id*. The same procedure happens with the remaining cores, until finally the core with

---

Algorithm 5.2: Gather array of length n distributed among p cores

---

```
// Assumes p = 2^i and n mod p = 0
for i ← 0 to log p − 1 do
    if id mod 2^i = 0 then
        if id/2^i mod 2 = 1 then
            send array of length 2^i n/p to id − 2^i
        else
            receive array starting at position (id + 2^i)n/p from id + 2^i
        end
    end
end
```

$id = 0$ has the result, as detailed in Algorithm 5.2. There are $\log p$ steps and the message size at step i is $2^i n/p$, therefore the communication time is:

$$t_{\text{gather}} = \lambda \log p + n(p-1)/(p\beta) = O(n + \log p).$$

We can now compare the two versions of matrix-vector multiply. The 1D agglomeration has complexity

$$t_{1D} = O(n^2/p + n \log p)$$

as the broadcast has a higher complexity than the gather. Note that for this algorithm to be cost efficient n must be greater than $p \log p$. The 2D agglomeration requires $\sqrt{p}$ concurrent reductions of arrays of length $n/\sqrt{p}$ among $\sqrt{p}$ cores (Figure 4.14) and $\sqrt{p}$ concurrent broadcasts of arrays of length $n/\sqrt{p}$ among $\sqrt{p} + 1$ cores (except for the first block of y, which is broadcast among $\sqrt{p}$ cores; see Figure 4.15). This results in an overall complexity of:

$$
\begin{aligned}
t_{2D} &= O(n^2/p + (n/\sqrt{p}) \log \sqrt{p} + (n/\sqrt{p}) \log \sqrt{p}) \\
&= O(n^2/p + (n/\sqrt{p}) \log p).
\end{aligned}
$$

The two dimensional version has less communication overhead by a factor of $\sqrt{p}$ than the one dimensional version, and therefore has better scalability.

5.3 BARRIERS TO PERFORMANCE

Achieving good performance is the goal of parallel computing, but it can be frustrating. Even if a parallel programmer has a reasonable performance goal of sub-linear speedup, initial performance results are often far below expectations. Understanding the most common barriers to performance can greatly reduce development time. Barriers to performance can be found at both the algorithmic and implementation levels.

Load Imbalance

Load imbalance is a fundamental concern in parallel computing, and one we have introduced in Section 4.1. Linear speedup is only possible if the work is shared equally among all cores, assuming the cores have the same processing power. If there is load imbalance, then some of the cores will be idle waiting for the rest of the cores to complete, and the overall execution time will be greater than if all the cores receive an equal share of work (see Figure 4.1).

Decomposition often results in tasks that have an unequal workload. The tasks in a parallel merge are a case in point, where the size of the sub-arrays to be merged is data-dependent. Another example is fractal generation, where the computation time of pixels depends on how many iterations are required before divergence or the threshold is reached. It's also possible that the load is imbalanced due to unequal processing power of cores. This can occur when the parallel computer contains cores with different processing power, or when time-sharing with other processes causes the processing power of cores to vary significantly. If the load imbalance is predictable then it can be dealt with in advance. For example, in the case of a computer with unequal cores, each core can be assigned work proportional to its processing power. More frequently, however, the imbalance is data dependent, and is therefore unpredictable in advance.

Strategies for alleviating load imbalance include cyclic allocation of tasks to cores, as we saw with parallel loops (Section 4.4). This is an example of *static load balancing*. *Dynamic load balancing* can also be used, such as with the master-worker pattern. While dynamic load balancing can lead to better results, there is a price to be paid in the overhead associated with dynamically assigning work. An important choice to be made, whatever strategy is used, is the granularity of tasks. Its easy to see that the greater the number of tasks of different workloads, the better load balance can be achieved. There are a few disadvantages of having very fine grained tasks. One is that the data each core works on is more scattered in memory, than if tasks work on more coarse grained contiguous chunks of data. This can lead to poor cache locality. Another is that more tasks means more overhead when dynamic load balancing is used. In practice, this means that right granularity has to be found experimentally.

Communication Overhead

In an embarrassingly parallel program each core operates on its own local data and does not need to share data or communicate data with other cores. As we know, most parallel programs require access to remote data. For distributed parallel programs this means that communication is the largest barrier to performance. We have already seen with matrix-vector multiplication and the Game of Life how agglomeration affects the communication overhead. In both these examples a 2D agglomeration results in less communication than an 1D agglomeration.

Communication can be further reduced by combining messages. Compare sending two messages of size m to sending one message of size $2m$. The time for each option is $2(\lambda+m/\beta)$ and $\lambda + 2m/\beta$, respectively. Sending one message reduces the time by λ. Whether this is significant will depend on the magnitude of the latency and the size of the message relative to the bandwidth.

We can combine messages to potentially reduce the execution time of the Game of Life. Communication is required to keep the ghost elements up to date. The parallel Game of Life in Algorithm 4.16 requires communication every iteration to exchange boundary values with its neighbors. If we add an extra layer of ghost cells then communication is only required every two iterations. In even iterations elements in $(n/p) + 2$ rows are updated, which includes the inner ghost rows. Note that the update of the inner ghost cells is replicating the same updates taking place in a neighboring core. In odd iterations only the n/p non ghost rows are updated. The new algorithm is shown in Algorithm 5.3 and in Figure 5.7.

What we have done in Algorithm 5.3 is traded off redundant computation of $2n$ values

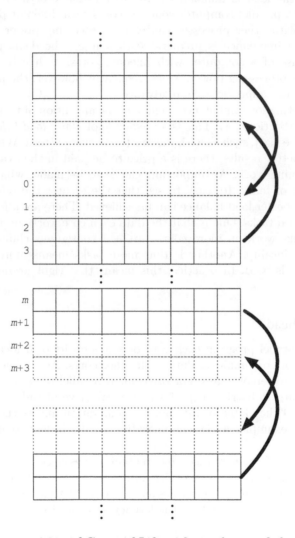

Figure 5.7: 1D decomposition of Game of Life with two layers of ghost cells. $m = n/p$.

Algorithm 5.3: Game of Life with Message Passing and an extra layer of ghost cells. Does not include initialization and display of grid. Changes made to Algorithm 4.16 shown in **bold**.

// each task has $(n/p + 4) \times n$ arrays $grid$ and $newGrid$

$nbDown \leftarrow (id + 1) \bmod p$
$nbUp \leftarrow (id - 1 + p) \bmod p$
$m \leftarrow n/p$ // Assume $n \bmod p = 0$
for $k \leftarrow 0$ **to** $N - 1$ **do**
    offset $\leftarrow k$ **mod 2**
    if offset $= 0$ **then**
        // nonblocking send of boundary values to neighbors
        nonblocking send $grid[\boldsymbol{m}..\boldsymbol{m} + \boldsymbol{1}, 0..n - 1]$ to $nbDown$
        nonblocking send $grid[\boldsymbol{2}..\boldsymbol{3}, 0..n - 1]$ to $nbUp$
        // receive boundary values from neighbors into ghost elements
        receive from $nbDown$ into $grid[\boldsymbol{m} + \boldsymbol{2}..\boldsymbol{m} + \boldsymbol{3}, 0..n - 1]$
        receive from $nbUp$ into $grid[\boldsymbol{0}..\boldsymbol{1}, 0..n - 1]$
    end

    foreach $cell\ at\ coordinate\ (i, j) \in (1 +$ offset$..\boldsymbol{m} + \boldsymbol{2} -$ offset$, \boldsymbol{0}..\boldsymbol{n})$ **do**
        updateGridCell($grid, newGrid, i, j$)
    end
    swap references to $newGrid$ and $grid$
end

---

every 2 iterations for a reduction in communication by 2λ. We can similarly add another layer of ghost elements to the 2D agglomeration. This type of communication/computation tradeoff is often encountered. The choice of whether to use redundant computation depends on the characteristics of the parallel computer. Experiments can reveal whether the redundant computation takes less time than the communication time saved.

An even more common technique to reduce communication is to hide it by overlapping it with computation. This can be done for the Game of Life by computing inner elements, which do not depend on the ghost elements, while the communication takes place. Starting again from Algorithm 4.16, we can move the calls to receive after the inner elements have been computed, as shown in Algorithm 5.4.

Recall that a non-blocking send returns immediately. This means that rows n/p and 1 of the $grid$ array should not be modified until it is safe to do so. There is no danger here since the $grid$ array isn't modified within each iteration. If the computation of the $(n/p - 2)n$ cells takes longer than the communication of $2n$ cells then the communication can be completely hidden.

Communication Overhead for Shared Memory

A naive view of shared memory programming would be that communication isn't an issue since there is no explicit communication and that all threads can access any memory location with the same latency. This view is incorrect because of the memory hierarchy, which we saw in Chapter 2. Data in cache memory is accessible in orders of magnitude less time than main memory. Distributed memory modules in NUMA systems mean that accessing remote data is done more slowly than with local data. Many programmers are already familiar

---

Algorithm 5.4: Game of Life with Message Passing with overlapping of communication and computation. Does not include initialization and display of grid. Changes made to Algorithm 4.16 shown in **bold**.

```
// each task has (n/p + 2) × n arrays grid and newGrid
```

$nbDown \leftarrow (id + 1) \bmod p$
$nbUp \leftarrow (id - 1 + p) \bmod p$
$m \leftarrow n/p$ // Assume $n \bmod p = 0$
for *a number of generations* **do**
    // nonblocking send of boundary values to neighbors
    nonblocking send $grid[m, 0..n - 1]$ to $nbDown$
    nonblocking send $grid[1, 0..n - 1]$ to $nbUp$

    foreach *cell at coordinate* $(i, j) \in (2..m - 1, 0..n)$ **do**
        updateGridCell($grid$, $newGrid$, i, j)
    end
    // receive boundary values from neighbors into ghost elements
    receive from $nbDown$ **into** $grid[m + 1, 0..n - 1]$
    receive from $nbUp$ **into** $grid[0, 0..n - 1]$
    foreach *cell at coordinate* $(i, j) \in (1, 0..n) \wedge (m, 0..n)$ **do**
        updateGridCell($grid$, $newGrid$, i, j)
    end
    swap references to $newGrid$ and $grid$
end

---

with optimizing programs to increase locality and hence re-use data in the cache as much as possible.

Even though there is no explicit communication in shared memory programs, communication between cores and memory modules occurs each time a thread loads and stores data. Even if threads only work on their own data, that data must be loaded into the cache memory associated with the core where the thread is running. There is also communication when threads share data. For example in the Game of Life, when a thread updates an element on its inner boundary it will need to access an element that belongs to another thread. As a result shared memory implementations with 1D and 2D agglomerations will result in the same $O(n)$ and $O(n/\sqrt{p})$ communication time, respectively, as the distributed memory implementations [76].

Shared memory programs do have the important property that they don't have to redistribute data, as was the case for the distributed matrix-vector multiplication algorithms. Because of this property, the 1D matrix-vector shared memory multiplication algorithm is more efficient than the 2D algorithm, since no sharing of data between threads is required in the former case. This is in contrast with the distributed algorithm, which for the 1D version required a broadcast of the resulting vector (see Section 5.2.2).

False Sharing

As we discussed in Chapter 2, cache memory enables increased performance without the explicit involvement of the programmer. Cache coherent protocols make the cache almost disappear for parallel programmers. The reason it isn't eliminated as a concern is that cache coherence is maintained for each *cache block*, which contains multiple data elements. If cores

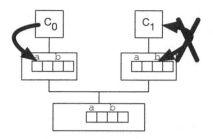

Figure 5.8: False Sharing. Core C_0 is writing to a, which prevents C_1 from accessing b since they both are in the same cache block.

are accessing distinct data (a and b in Figure 5.8) that are in the same cache block in their local cache they cannot do concurrent writes (or writes and reads). The programmer thinks the threads are making concurrent accesses to these data elements, but in fact they cannot if the block is in coherent caches. This is called *false sharing*, since the intent isn't to share a and b, but the sharing is forced because they lie on the same cache block. While false sharing does not affect correctness it does reduce performance.

For example, recall the reduction in a shared loop in Algorithm 4.10, which had each thread accumulate partial sums in the elements of an array. While this would execute correctly it would suffer from false sharing. Depending on the number of threads and the cache block size several elements of the *psum* array could reside in a single block and some writes to $psum[id]$ would be serialized. False sharing does not occur if the data being accessed is in a single cache that is shared by multiple cores.

Hierarchical Algorithms

Another approach to achieving locality and hence reducing communication is to create hierarchical algorithms to match hierarchical computing platforms. CUDA or OpenCL programs will often be hierarchical in nature, with thread groups working independently at each level. This reduces the need for global synchronization and results in good locality with each thread group working on its own data. We saw one example of this with the GPU reduction in Algorithm 4.14.

A hierarchical approach is also worth considering for distributed memory algorithms. Since cores in a node of a cluster can share memory, it is possible to apply a coarser grained agglomeration with one task per node rather than one task per core. This task in the distributed memory task graph can then be decomposed into tasks that share memory. For example, for the 2D Game of Life executing on p nodes, each node could compute the elements in its block by using a shared memory algorithm, with one core taking on the task of exchanging boundary elements with other nodes. Another example is merge sort. Instead of each core performing a sequential sort of n/p elements, each group of cores on a shared memory node can perform a parallel shared memory sort, followed by the same distributed merge. This approach has been often seen in the form of hybrid MPI/OpenMP programs [20].

Inherently Sequential Execution

It is often the case that not all computations can be done in parallel. In the parallel k-means algorithm the computation of the coordinate of new cluster centers, by dividing the sum of coordinates of vectors assigned to each center by the number of vectors, is done by one task. In the fractal computation or the Game of Life, the output needs to be saved to disk or displayed. If output is done by a single core, which is often the case, then the overall speedup will be limited. The argument can be made that for a fixed problem size, a sequential portion will limit speedup as the number of processors increases. This observation is known as *Amdahl's law*:

$$\text{speedup} = \frac{t_1}{ft_1 + (1 - f)t_1/p}$$

where f is the fraction of the sequential execution time t_1 of the operations that cannot be done in parallel. In the limit where $p \to \infty$ this gives an upper bound on the speedup of $1/f$. Note that if $f = 0$, Amdahl's law predicts a perfect speedup of p. It doesn't take into account any barriers to performance other than sequential computation.

In practice, this barrier to performance is not significant for properly designed parallel programs. Parallel file systems remove the I/O bottleneck for applications operating on large data sets. The sequential portion often has a lower complexity that the parallel portion. For example, in k-means the sequential computation is only done once each iteration for each cluster, and does not depend on the problem size. Since more cores are usually used to solve larger problems (see weak scaling discussion above) the sequential fraction is not a constant but instead decreases as the problem size is increased.

Gene Amdahl

Amdahl's law is frequently cited in introductions to parallel programming, but its true relevance is to computer architecture. The originator of the law made a huge contribution to the history of computing. In the early 1960s he was the chief architect of the IBM System/360, which began the corporation's most profitable mainframe product line. This was the first time the term *architecture* was applied to a computer design. The System/360 did not use parallel processing, but a rival computer, the ILIAC IV was a SIMD design with 64 processors. In 1967, Gene Amdahl argued at the AFIPS Spring Joint Computer Conference that because the operating system of the ILIAC IV took 25 to 45 percent of the machine cycles, parallel programs could only achieve a speedup of 2 to 4. His observation was that there would need to be a very high fraction of instructions from a single instruction stream that could be executed in parallel in order to observe significant speedup, and that this wasn't the case in practice. His argument, which was referred to later as Amdahl's law, is still relevant to processor designers, because they have to take into account the inherently sequential nature of many of the instructions that processors execute. Amdahl's law is not an expression of pessimism about parallel computing, because successful parallel computers have multiple instruction streams, and as John Gustafson noted in 1988, the effect of increasing problem size diminishes the importance of inherently sequential instructions. Gene Amdahl isn't only famous for his law or for the design of the System/360. He also invented the idea of the computer clone when he formed Amdahl corporation in 1970. Amdahl mainframes had an IBM-compatible instruction set, and successfully competed with IBM's mainframes, capturing 20% of the market.

Amdahl's law is relevant when optimizing a sequential program, whether through sequential optimization techniques or through incremental parallelization. Even if 90% of the program, measured in execution time, can be parallelized, the speedup will still be limited to 10. This is a serious limitation of incremental parallelization. A highly scalable parallel program requires redesign of the algorithms and a new implementation.

Memory

Memory can be a barrier to performance when there is not enough of it or when multiple threads contend for access to the same data. If memory capacity is a problem the decomposition may have to be be more fine grained to allow the data for tasks executing in a multiprocessor to fit in the available memory. False sharing, which occurs when multiple threads are accessing data in a single cache block, is an example of contention for shared memory leading to reduced performance.

More generally, the contention produced by multiple threads sharing a data structure needs to be carefully addressed to ensure both correctness and performance. For instance, to implement a shared memory parallel program for the histogram problem of Chapter 3 one might decide to use a single shared array for the histogram (see Section 4.4.2). Access by threads to the histogram would need to be protected by a critical section to ensure correctness. As more threads were used this would lead to significant slowdown as threads would often be idle waiting for access to the histogram. There is a large body of knowledge on working with concurrent data structures in order to ensure correctness and performance, which applies not only parallel programs but to all situations where data is shared. For parallel programming this problem of contention can be avoided by using private data structures as much as possible, and aggregating the results.

Contention can't be entirely reduced for shared memory programs, however. Even if threads only work on private data, they all need to access memory. Concurrent access to data can overwhelm the memory interconnection network and limit the available bandwidth. This problem can be mitigated by migrating threads or processes in a multiprocessor to increase throughput; however, this is normally done at the system level and not by the programmer. It can sometimes be influenced by the programmer in the selection of a mapping policy, such as the placement of processes in nodes of a cluster. For instance, the resource management system for a cluster may offer the choice of scattering processes across multiprocessor nodes on the one hand, or filling up nodes one at a time on the other.

Other Overhead

There is usually overhead involved in implementing a parallel algorithm, some avoidable, some not. As an example of unavoidable overhead, if one needs to send data that is not contiguous in memory in a single message then it will have to be copied into a buffer first. Or a thread may need to calculate the starting and ending index of the data it is responsible for. There are plenty of opportunities for avoidable overhead. For example, for the fork-join merge sort in Algorithm 4.6 it would have been simpler for each merge task to allocate a temporary array, rather than have merges alternate between two pre-allocated arrays. Allocating extra memory wastes space and time. The same practices that lead to efficient sequential programming are also relevant to parallel programming.

5.4 MEASURING AND REPORTING PERFORMANCE

Measuring performance properly is crucial when tuning a program and when reporting results in the literature. During development good use should be made of performance profiling tools. These tools can be vital to identifying bottlenecks in execution. We'll focus here on taking measurements for publication.

Timing

The execution time of a parallel program is usually measured in wall-clock time. It's best measured by inserting calls to timing functions. It's important to know the resolution of the timer being used, as the reported time will not be meaningful if it is less than the resolution. Where to place timers is an important decision. There may be good arguments for leaving out reading input files from disk and writing final results to disk. It may also be valid to leave out the initial distribution of data. It all depends on what one wishes to measure. It is very important to clearly indicate what is being left out when reporting timing results.

Data Size

The choice of problem parameters and input data sets also needs to be carefully explained, just as when reporting the performance of sequential programs. An important consideration is the choice of a range of data set sizes, in order to show how performance varies with problem size. Results can be reported with multiple strong scaling speedup plots and/or a single weak scaling plot.

Experimental Parameters

When publishing results it is crucial to give all experimental details of the application (data sets, parameters, ...) and the computer. The processor models, interconnect networks, operating system and compiler should all be clearly identified, including system software version numbers.

Repeated Experiments

Ideally, one would be able to run a single experiment for each instance of the test data set. Unfortunately there are a number of reasons why multiple runs can produce different execution times. One's program is sharing the computing resources with operating system processes. This problem has been exacerbated by the use of commodity software in most parallel computers. A computer dedicated to running a single program at a time could make do with a much more lightweight operating system.

For shared memory programs, each run can lead to a different order of accesses by threads to shared memory, which can result in different idle times and hence variation in execution time. Also, the dynamic allocation of tasks to threads can vary from run to run. All this means that you're likely to find some variability when you run your program more than once. Therefore multiple runs need to be done for each problem instance, and either the median or mean should be reported. When there are outliers that have times much greater than the rest, the median provides a better indication of the expected execution time.

5.5 FURTHER READING

Work-depth analysis is discussed in the chapter on multithreaded execution in the third edition of Cormen et al.'s *Introduction to Algorithms* [13]. It is also used in Blelloch and Maggs's chapter on parallel algorithms in the *Algorithms and Theory of Computation Handbook* [9]. Performance metrics not discussed in this chapter include the *isoefficiency metric*, which refers to how much problem size has to grow to maintain efficiency. This metric is discussed in Grama et al.'s *Introduction to Parallel Computing* [30]. A discussion of the two scan algorithms referred to in this chapter, in the context of GPU programming, can be found in Harris's *Parallel Prefix Sum (Scan) with CUDA* [34]. A nice example of what's involved in implementing collective communication routines is found in Thakur and Gropp's *Improving the Performance of Collective Operations in MPICH* [72]. There have been many reevaluations of Amdahl's law since it was first published [2]. The most significant one considered the scaling of problem size with number of processors, and led to Gustafson's law [33]. Another is Yavits and Ginosar's *Effect of communication and synchronization on Amdahl's law in multicore systems* [76], which exposes the communication cost of shared memory operations. There are many investigations into the effect of thread and process placement on performance, with two good examples being Majo and Gross's *Memory System Performance in a NUMA Multicore Multiprocessor* [48] and Jeannot et al.'s *Process Placement in Multicore Clusters* [42]. Just as you shouldn't learn to drive a car by watching other drivers, you also shouldn't learn how to report performance results from papers in the literature. This is part of the larger issue of reproducibility in computational research, which has been addressed by David Bailey and Victoria Stodden, among others [4].

5.6 EXERCISES

5.1 Analyze the work and depth of the pointer jumping algorithm described in Section 3.7.

5.2 A simple shared memory algorithm was described in Section 4.4.2 for counting duplicates of an array of elements that have a limited range of nonnegative values. Design a distributed memory algorithm to solve this problem and analyze its performance.

5.3 Implement the algorithm in Question 5.2 using MPI, and compare the experimental scalability with the predicted performance.

5.4 What is the complexity of the parallel runtime and the cost of the pipelined merge sort Algorithm 3.4? Is it cost optimal? How does it compare with the fork-join merge sort Algorithm 4.6?

5.5 Implement the SPMD scan of Algorithm 5.1, comparing the use of the two parallel scan algorithms. Discuss why or why not the performance of this coarse grained scan depends significantly on the scan algorithm used.

5.6 Analyze the performance of the distributed memory subset sum Algorithm 4.19. By how much does problem size need to grow with p to maintain efficiency?

5.7 The subset sum Algorithm 4.11 is based on shared memory tasks with dependencies. Rewrite this algorithm using the same block wavefront decomposition, but for a message passing platform. Analyze the expected performance.

5.8 Implement the algorithm in Question 5.7 using MPI, and compare the experimental scalability with the predicted performance.

5.9 Write a message passing Floyd-Warshall algorithm (see Exercise 3.9) and analyze its performance. By how much does problem size need to grow with p to maintain efficiency?

5.10 Implement the algorithm in Question 5.9 using MPI, and compare the experimental scalability with the predicted performance.

5.11 Experimentally study the suggested improvements to the Game of Life (Algorithms 5.3 and 5.4). First, experimentally determine the time required to update a row of the grid. Then determine the experimental latency and bandwidth using a simple benchmark that sends messages of different sizes back and forth. Predict whether the suggested improvements will work for your computer. Then test the full programs for each alternate algorithm to verify your prediction.

CHAPTER 6

Single Source Shortest Path

Graphs are useful for modeling many natural and artificial phenomena. In a graph representing a transportation network, the vertices represent cities and the edges indicate which cities are directly connected by transportation links. The edge weights represent either distance or time to travel between cities. In a social network graph the vertices represent individuals and the edges represent direct relationships between individuals. Graphs representing real-world networks can reach sizes of billions of vertices. This means that graph analysis can require parallel computing to provide sufficient memory and computing power.

The distance between vertices is an important property of graphs, which is given by the weight of the shortest path between the vertices. In transportation networks it indicates the distance or time required to travel between cities. Distance measures the degree of separation between two individuals in a social network.

There are three related shortest path problems for graphs: between all pairs of vertices, between one vertex and all others (single source), and between 2 vertices. In this chapter we'll focus on the *single source shortest path* (SSSP) problem. It's also possible to find the shortest path between two vertices using an SSSP algorithm and stopping when the final distance to the destination vertex has been found.

Consider the graph in Figure 6.1, where the edges are weighted. The distances between vertex 0 and vertices $1 - 7$ are $\{2, 1, 4, 6, 5, 7, 8\}$. Even though vertices 0 and 1 are directly connected, the shortest path between them passes through vertex 2. This path is part of the shortest path to vertices 3,5,6,7.

Algorithms for solving the SSSP can be specialized for particular types of graphs, such as road networks. We'll consider algorithms suitable for all graphs with non-negative real-

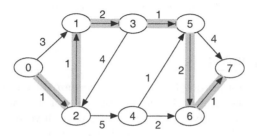

Figure 6.1: A weighted graph, with vertex labels and edge weights. The shortest path between vertices 0 and 7 is highlighted.

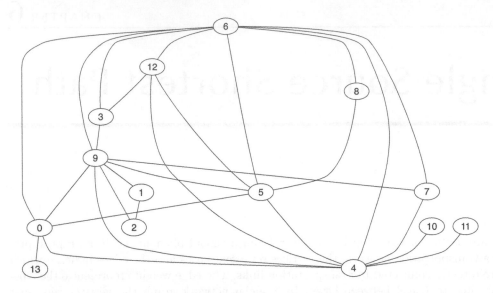

Figure 6.2: Undirected graph with vertices of widely different degree.

valued edge weights. Edges in undirected graphs are considered to be pairs of directed edges.

The performance of SSSP algorithms is sensitive to the properties of the graph being analyzed, in particular the *degree distribution* and *diameter*. The cities in a road network are directly connected to a small number of other cities, so the degrees of the vertices in the graph are low and of similar magnitude. In contrast, an airline network includes a few hub airports that are directly connected to many other airports, so the graph has both low and very high degree vertices. For example, consider the graph in Figure 6.2. More than half the vertices have degree less than 4, while vertices 4, 6, and 9 have degree 8. Graphs like this are often found in real-world networks. They are called *scale-free* when their degree distribution follows a power law, that is, the number of vertices of degree d is $O(d^{-\lambda})$, where λ is a small constant.

The diameter of a graph is given by the length of the longest shortest path between all pairs of vertices, where distances are measured in hops (that is, ignoring weights). The diameter of the graph in Figure 6.1 is formally infinite, since not all vertices are connected. However, we can say that its diameter is 4 if we only consider paths of finite length, since the longest shortest path takes 4 hops (*e.g.* 0-1-3-5-7). The diameter of the graph in Figure 6.2 is 3, where paths can make use of high degree vertices to reach their destination in a few hops.

6.1 SEQUENTIAL ALGORITHMS

Several sequential algorithms for finding the single-source shortest paths follow the iterative labeling pattern outlined in Algorithm 6.1 [39]. The distance of the shortest paths is output in a distance array D, and array P gives the predecessor to each vertex in the shortest path that reaches the vertex. Array P allows the vertices in each shortest path to be enumerated. In the following algorithms we'll drop computation of P. The distance array is initialized to a value larger than the maximum possible distance, except for element s, which is set to 0. The core of all shortest path algorithms is the *relaxation* (lines 9–13), where edge e_{ij}

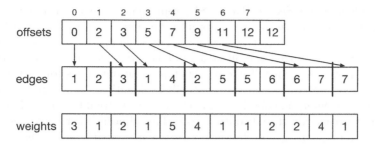

Figure 6.3: Compressed Sparse Row Storage of graph in Figure 6.1.

is considered and the distance to vertex j is updated if the path to j through the edge e_{ij} is shorter than the current distance $D[j]$. A list is used to store vertices that are being considered, beginning with the source vertex. Algorithms differ in how they implement this list and whether they use multiple lists. In each iteration of *label-setting* algorithms the distance value of a single vertex is permanently set. *Label-correcting* algorithms may update vertex distances multiple times, and vertices may be removed and added to the list multiple times as well.

Algorithm 6.1: SSSP algorithm structure using labelling.

Input: Graph with vertices V ($|V| = n$) and edges E with weights C, source vertex s.
Output: D: distance between s and all other vertices. P: pointer to predecessor to each vertex in shortest path.

1: **for** $i \leftarrow 0$ *to* $n-1$ **do**
2:     $D[i] \leftarrow \infty$
3: **end**
4: $D[s] \leftarrow 0$
5: list $\leftarrow \{s\}$
6: **while** list $\neq \emptyset$ **do**
7:     remove vertex i from list
    `// Relax each out-edge of vertex i`
8:     **foreach** *edge* $e_{ij} \in E$ **do**
9:         **if** $D[j] > D[i] + c_{ij}$ **then**
10:             $D[j] \leftarrow D[i] + c_{ij}$
11:             $P[j] \leftarrow i$
12:             **if** $j \notin$ list **then** list $\leftarrow$ list $\cup \{j\}$
13:         **end**
14:     **end**
15: **end**

6.1.1   Data Structures

The two standard graph representations, adjacency matrix and adjacency list, are well known [13]. Adjacency lists are usually recommended, as most graphs encountered in practice are *sparse*, that is they have much fewer than the maximum number of edges ($|V||V-1|$). However, there are other representations that can be borrowed from sparse linear algebra.

In particular, the *Compressed Sparse Row* (CSR) representation uses two arrays: an array of edges and an offset array of size $|V|$ to index into the edge array, as shown in Figure 6.3. An additional array is used to store edge weights. The edge array contains the adjacency lists stored contiguously. Note that the offset array has one extra value at the end to store the size of the edge array. We can tell if row i is empty (no out-edges for vertex i), as the i-th and $(i + 1)$-th elements of the offset array will be identical. The fact that edges are stored contiguously in memory means that CSR is more cache-friendly than adjacency lists, whose nodes are not necessarily stored contiguously. The adjacency list representation more easily allows one to modify the graph, however this is not needed to solve the SSSP.

6.1.2 Bellman-Ford Algorithm

The standard Bellman-Ford algorithm is label-correcting to the extreme (Algorithm 6.2), as all edges are relaxed in each iteration. This algorithm has two advantages: it can handle negative weights, as long as it checks for negative-weight cycles, and it provides a high degree of concurrency, as we'll see in the next section. This form of the algorithm is not used in practice, as many of the iterations are unnecessary. Edges adjacent to vertices whose distance values haven't changed from previous iterations don't need to be relaxed. A practical Bellman-Ford algorithm follows Algorithm 6.1, where vertices are placed into a list, starting with the source vertex. The order in which vertices are removed is not important, although a sequential implementation can use a first-in-first-out (FIFO) queue. Instead of relaxing each edge, this modified Bellman-Ford algorithm only relaxes edges from vertices whose distances have been updated.

Algorithm 6.2: Standard Bellman-Ford, for nonnegative weights.

Initialize D
for $k \leftarrow 1$ *to* $n - 1$ **do**
    foreach *edge* $e_{ij} \in E$ **do**
        if $D[j] > D[i] + c_{ij}$ **then**
            $D[j] \leftarrow D[i] + c_{ij}$
        end
    end
end

Table 6.1 traces the modified Bellman-Ford algorithm for the graph in Figure 6.1, showing the queue after each iteration of the `while` loop, with the current distance to each vertex shown in parenthesis, and the vertices with updated distances identified. We can see that some vertices are inserted and removed from the list multiple times, namely vertices 1 and 3. The modified Bellman-Ford algorithm has the same $O(|V||E|)$ time complexity as the original algorithm, but in practice results in much better performance for sparse graphs. Lower diameter graphs tend to require fewer iterations, since the paths are shorter, so fewer iterations are needed to propagate updates.

6.1.3 Dijkstra's Algorithm

Dijkstra's algorithm updates each vertex once by only relaxing edges from vertices that have reached their final distance. It does this by selecting the vertex with the minimum

Table 6.1: Trace of Modified Bellman-Ford algorithm for Figure 6.1.

Queue	vertices updated
0(0)	
1(3), 2(1)	1, 2
2(1), 3(5)	3
3(5), 1(2), 4(6)	1, 4
1(2), 4(6), 5(6)	5
4(6), 5(6), 3(4)	3
5(6), 3(4), 6(8)	6
3(4), 6(8), 7(10)	7
6(8), 7(10), 5(5)	5
7(9), 5(5)	7
5(5)	
6(7)	6
7(8)	7

Table 6.2: Trace of Dijkstra's algorithm for Figure 6.1.

Queue	vertices updated
0(0)	
2(1), 1(3)	1, 2
1(2), 4(6)	1, 4
3(4), 4(6)	3
5(5), 4(6)	5
4(6), 6(7), 7(9)	6, 7
6(7), 7(9)	
7(8)	7

distance at each iteration. It can be expressed by Algorithm 6.1 with a min-priority queue instead of a list. Table 6.2 traces Dijkstra's algorithm for the graph in Figure 6.1.

Compared with the Bellman-Ford example, there are fewer iterations and distance updates. Each vertex is inserted and removed from the queue only once. Since each vertex is only removed once, the number of iterations of the outer loop is $|V|$ and the total number of iterations of the inner foreach loop is $|E|$. If we use a binary heap for the priority queue, which takes $O(\log|V|)$ to update vertices and remove the minimum vertex, the complexity of Dijkstra's algorithm is $O((|V| + |E|)\log|V|)$. Even though the Bellman-Ford algorithm has a much higher run time in the worst case than Dijkstra's algorithm, in practice it doesn't usually reach the worst case and may even outperform the latter algorithm. Unfortunately we don't know in the average case how Bellman-Ford will perform.

6.1.4 Delta-Stepping Algorithm

Meyer and Sanders's delta-stepping is a label-correcting algorithm that has average-case linear complexity [51]. It can also be decomposed into a parallel algorithm with better work efficiency than Belman-Ford, as we'll see in the next section. It uses an array of buckets to store vertices, where each bucket stores vertices that have distances within a range Δ.

Let's take the same example of Figure 6.1, and use an array B of buckets, with $\Delta = 5$. Bucket i holds vertices with distance in the range $[i\Delta, (i+1)\Delta)$. Each relaxation will place a vertex in one of the buckets, and may also move it from another bucket. We first place the source vertex in the first bucket, since is has distance 0. When we relax edges e_{01} and e_{02} vertices 1 and 2 are placed in the first bucket since their distances are less than 5. We then proceed by removing vertices from the first bucket and relaxing their out-edges. We keep doing this until there are no more vertices in the first bucket, then proceed in the same manner through the other buckets, as shown in Table 6.3.

Table 6.3: Trace of Delta-Stepping algorithm for Figure 6.1.

B[0]	B[1]	vertices updated
0(0)		
1(3), 2(1)		1, 2
2(1)	3(5)	3
1(2)	3(5), 4(6)	1, 4
3(4)	4(6)	3
	4(6), 5(5)	5
	5(5), 6(8)	6
	6(7), 7(9)	6, 7
	7(8)	7

Note that vertices can be inserted into a bucket multiple times, as is the case for vertex 1. Also, vertices can move between buckets, as for vertex 3, which moved from the second to the first bucket. Basically, what we are doing is using the modified Bellman-Ford algorithm, but using multiple queues, and we're completely emptying one queue before we start processing vertices in the next queue. This algorithm retains part of the benefit of Dijkstra's algorithm, as illustrated in this example for vertex 5. Since this vertex is placed in the second bucket, it isn't processed until edges from vertex 3, which is the last vertex in the first bucket, are relaxed. This means that this vertex gets its final distance in one update, whereas the Bellman-Ford algorithm required two updates.

Ideally we would be able to only do relaxations that place vertices in the same bucket. This would have saved us from updating vertex 3 twice, since it wouldn't have been placed in the second bucket edge in the third iteration above. While this isn't possible, Meyer and Sanders's algorithm is able to reduce the number of relaxations that place vertices in other buckets by classifying edges into light (weight $\leq \Delta$) and heavy (weight $> \Delta$) categories. Clearly, relaxing a heavy edge must place the vertex it points to in another bucket. The delta stepping algorithm, given in Algorithm 6.3 [47, 51], first iteratively relaxes light edges from a bucket, before relaxing the heavy edges.

A preprocessing step (lines 1–5) places all the edges into either a light or a heavy set. Light edges from the vertices in bucket i are relaxed, and these vertices are removed from the bucket. These relaxations may reinsert vertices back in bucket i, so the process iterates until the bucket is empty (lines 10–17). The final step in the process for bucket i is to relax heavy edges from the vertices for which light edges were relaxed, therefore they need to be remembered (line 12). The sets of vertices terminating edges and candidate weights are stored in request sets Req (lines 11 and 18). The `relax` procedure performs the relaxation, which, if accepted, moves the vertex into the appropriate bucket.

The Δ parameter can be tuned to make the algorithm closer to Dijstra's algorithm ($\Delta \to 0$) or closer to Bellman-Ford's algorithm ($\Delta \to \infty$). Meyer and Sanders proved that delta-stepping takes $O(|V| + |E| + d \cdot L)$ time on average for graphs with random weights, where d is the maximum degree and L is the maximum shortest path weight from the source vertex, and $\Delta = O(1/d)$.

Algorithm 6.3: Meyer and Sanders's Delta-Stepping Algorithm

Input: Graph with vertices V ($|V| = n$) and edges E with weights C, source vertex s.
Output: D: distance between s and all other vertices.

1: **foreach** *vertex* $v \in V$ **do** //classify edges as light or heavy
2:     $H[v] \leftarrow \{e_{vw} \in E \mid c_{vw} > \Delta\}$
3:     $L[v] \leftarrow \{e_{vw} \in E \mid c_{vw} \leq \Delta\}$
4:     $D[v] \leftarrow \infty$
5: **end**
6: relax(s, 0)// places s in first bucket ($B[0]$)
7: $i \leftarrow 0$
8: **while** B *is not empty* **do**
9:     $R \leftarrow \emptyset$
10:    **while** $B[i] \neq \emptyset$ **do**
11:        $Req \leftarrow \{(w, D[v] + c_{vw}) \mid v \in B[i], e_{vw} \in L[v]\}$
12:        $R \leftarrow R \cup B[i]$// remember vertices deleted from bucket
13:        $B[i] \leftarrow \emptyset$
14:        **foreach** $(w, x) \in Req$ **do**
15:            relax (w, x)
16:        **end**
17:    **end**
18:    $Req \leftarrow \{(w, D[v] + c_{vw}) \mid v \in R, e_{vw} \in H[v]\}$
19:    **foreach** $(w, x) \in Req$ **do**
20:        relax(w, x)
21:    **end**
22:    $i \leftarrow i + 1$
23: **end**

    // Relax edge to vertex w with candidate weight x
    // If accepted assign to appropriate bucket
24: **Procedure** relax(w, x)
25:    **if** $x < D[w]$ **then**
26:        $B[\lfloor D[w]/\Delta \rfloor] \leftarrow B[\lfloor D[w]/\Delta \rfloor] \setminus \{w\}$
27:        $B[\lfloor x/\Delta \rfloor] \leftarrow B[\lfloor x/\Delta \rfloor] \cup \{w\}$
28:        $D[w] \leftarrow x$
29:    **end**
30: **end**

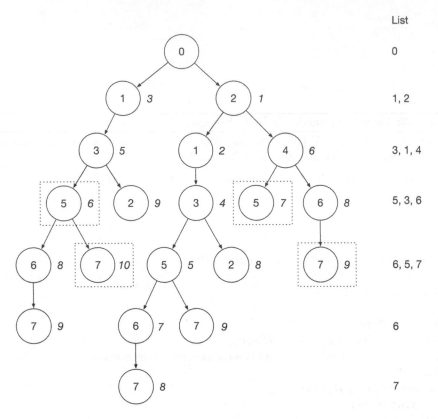

Figure 6.4: Task graph for Bellman-Ford algorithm for Figure 6.1, where tasks wait for all to complete at each level before moving to the next level. The dotted lines indicate race conditions on two of the levels. Labels in italics indicate a *candidate* distance to the vertex represented by the task. The contents of the list at each level is also shown.

6.2   PARALLEL DESIGN EXPLORATION

The fundamental task in SSSP algorithms is relaxation. Additional bookkeeping tasks may be required, such as constructing request sets and maintaining buckets in delta-stepping. Therefore, one approach to decomposition is to identify as many independent relaxation tasks as possible. An alternative approach is to perform a data decomposition of the vertices and/or the edges, and to associate a task with each subdomain. We'll consider task decomposition first.

Task Decomposition

There are two possibilities for decomposing the tasks in Algorithm 6.1 (Bellman-Ford). The most obvious one is to decompose the iterations of the `foreach` loop, where edges adjacent to a vertex are relaxed. These iterations are independent, but care must be taken when adding target vertices to the list (line 12), which requires suitable parallel set operations. The second type of decomposition is to process the contents of the list in parallel. The two decompositions can be combined to relax all edges out of all vertices in the list in parallel. Of course, each task in turn creates more tasks as a result of the relaxation, so the list will be repopulated. This process continues iteratively until the list is empty. Figure 6.4

illustrates a task graph for the problem in Figure 6.1, where all tasks on each level of the tree are completed before the tasks on the next level begin. A task represents a relaxation, and has child tasks when one or more relaxations are accepted and result in vertex updates. It's important to observe that this decomposition results in race conditions. In our example there are competing updates to vertex 5 and to vertex 7 at the fourth and fifth levels, respectively. Therefore the vertex updates need to take place atomically.

If we decompose the tasks in Dijkstra's algorithm, where the vertex that has the minimum distance is removed at each iteration of the `while` loop, we only have independent tasks for relaxations of edges from one vertex. The parallelism of this decomposition is $O(|E|/|V|)$, since the work is the same as the sequential algorithm and the depth is $|V| \log |V|$. This only provides significant parallelism if the graph is dense, where $|E|/|V|$ is not much less than $|V|$. Sparse graphs are more commonly used, and can have $|E| = O(|V|)$, which severely limits the parallelism.

The parallelism of the Bellman-Ford based decomposition depends on the characteristics of the graph. In the worst case, $|V|$ iterations are required, so the depth is $O(|V|)$, the work is $O(|E||V|)$, and therefore the parallelism is $O(|E|)$. Even in the average case, this decomposition provides much greater parallelism that the one based on Dijkstra's algorithm. The number of tasks available at each level depends on the degree distribution and diameter of the graph. It's easy to see that more relaxation tasks can be executed in parallel when the degree of the vertices is high. High diameter graphs have a lower number of active vertices than low diameter graphs, and the number of levels of the task graph is bounded by the diameter. Therefore low diameter graphs result in fewer steps with more tasks at each step.

To appreciate the effect of diameter consider the shortest paths from vertex 13 in Figure 6.2. If we construct a task graph like that of Figure 6.4, it results in a tree with four levels, with 5 vertices active at the third level (from the neighbors of vertex 0) and all vertices active at the fourth level. For another example consider two cartesian mesh graphs, where each vertex has an edge with 4 neighbors: one laid out in a $\sqrt{n} \times \sqrt{n}$ grid, and another laid out in a $n/32 \times 32$ grid. The first graph has a diameter of $2\sqrt{n}$ whereas the second has a diameter close to n. Both graphs have the same size, but the number of active vertices is much higher in the square graph.

This good parallelism comes at the price of work-inefficient execution, as the Bellman-Ford work is much higher in complexity than the Dijkstra work ($O(|E||V|)$ vs. $O((|V| + |E|) \log |V|)$). The delta-stepping algorithm allows us to tune the work done through the Δ parameter in order to balance parallelism and work-efficiency. The best tradeoff between the two, for random graphs, is expected to be for $\Delta = O(1/d)$. With this value of Δ the expected work is $O(d|V|)$. The expected number of iterations of the outer `while` loop over all buckets is expected to be $O(dL \cdot \frac{\log n}{\log \log n})$ [47, 51]. If tasks at each step can all execute in parallel, this is also the expected depth of a parallel implementation, which results in a parallelism of $O(\frac{n}{L} \frac{\log \log n}{\log n})$. This clearly shows that lower diameter (which means a lower L) graphs result in more parallelism.

Relaxation in delta-stepping Algorithm 6.3 is divided into two tasks: preparing request sets, and performing the relaxations. The latter task differs from relaxation in the other two sequential algorithms. Here it involves placing a vertex in a bucket and possibly removing it from another bucket. As with the other algorithms, care must be taken to eliminate race conditions, and parallel set operations are required if tasks are to execute in parallel.

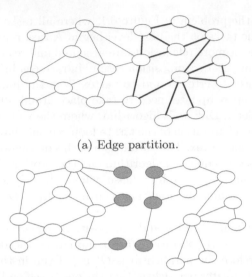

(a) Edge partition.

(b) Subgraphs resulting from edge partition. Gray vertices on boundary replicated in both subgraphs.

Figure 6.5: Edge partition example.

Data Decomposition

Decomposition of the problem graph is required for a distributed memory implementation. Since relaxations are performed across edges it makes sense to partition the edges of the graph, as illustrated in Figure 6.5. There are two types of vertices in each subgraph: 1) interior vertices, whose edges are all in the subgraph. 2) boundary vertices, which have one or more edges that belong to another subgraph. Boundary vertices are replicated in each subgraph that owns one of its edges.

The partition has a big impact on performance [38]. We've already seen this for the Game of Life in Chapter 5, and it will come up again for the Eikonal equation in Chapter 7. In those cases the best partitions produce subdomains of almost equal size (to balance the computation) and minimize the number of boundary vertices (to minimize communication). The situation of the SSSP is quite different. To see why, we need to look at the task that is associated with each subdomain.

A straightforward algorithm has each node compute the shortest paths for its subgraph, exchange distances for boundary vertices that were updated, and continue until all lists are empty [38]. Only distance information for boundary vertices needs to be communicated, since all shortest paths that cross boundaries pass through boundary vertices.

The solution will initially propagate through one or two (if the source is a boundary vertex) subgraphs. How soon other subgraphs are involved depends on the partition. This load imbalance may not be a problem if we need to find the SSSP from multiple sources. In that case each node can compute paths from all sources that pass through its subgraph.

After the first iteration, the amount of work involved depends on several characteristics of the partition. The total amount of communication depends on the number of boundary vertices, which is why graph partitioning for the purpose of data decomposition tries to minimize the number of boundary vertices. What's different about the SSSP problem is that the amount of computation varies not just with the size of the subgraph, but also with other characteristics such as the number of boundary vertices and the length of paths

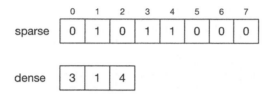

Figure 6.6: Sparse and dense representations of subset $\{3,1,4\}$ of set of 8 vertices. Dense array is unordered. Additional information for each vertex can be incorporated into a dense array or stored in a companion array.

from these vertices. The greater the number of boundary vertices the more information can be propagated between subgraphs, which reduces the number of iterations. The shorter the paths from these boundary vertices, the less computation is required to complete the resulting updates [38].

Graph partitioning is an NP-hard problem that has generated a large body of knowledge. Practical solutions use a heuristic method to minimize an objective function. A good partition for SSSP has subgraphs of roughly the same size, *maximizes* the number of boundary vertices and minimizes the maximum path length from boundary vertices. Minimizing communication is still important, and can be achieved by minimizing the number of interfaces between subgraphs. This means that we want each node to only communicate with a small number of other nodes.

6.3 PARALLEL ALGORITHMS

We'll discuss three parallel algorithms for three types of platforms: coarse-grained shared memory, SIMD on GPU, and distributed memory.

6.3.1 Shared Memory Delta-Stepping

We've seen that the decomposition of delta-stepping is quite straightforward, but we need to use efficient parallel set operations and eliminate race conditions. We need to build request sets and move vertices into buckets. Let's look at the set operations required for the first step.

Parallel Set Operations

We can represent subsets of $|V|$, which are needed for request sets, in a dense or sparse representation, or a combination of both, as shown in Figure 6.6. A dense representation stores the elements of the subset in a dense array, in no particular order, with each element containing the *id* of the vertex. A sparse representation uses an array of length $|V|$ with elements set to 1 identifying vertices in the subset. The dense representation allows us to efficiently visit elements of the subset, whereas the sparse representation allows a $O(1)$ query to check if a vertex is in the subset. The sparse representation also allows us to prevent duplicates from being stored in the subset.

Let's consider a couple of examples. First, let's use two threads to build a request set for Figure 6.1 starting from the second iteration of Table 6.3, where vertices 1 and 2 are in the first bucket. Thread 0 takes vertex 1 and locally stores a request $(3, 3+2)$, while thread 1 takes vertex 2 and locally stores requests $(1, 1+1)$ and $(4, 1+5)$. To create an array of

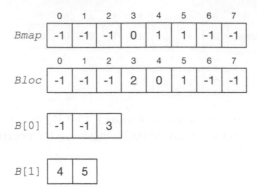

Figure 6.7: Bucket data structure, shown for three vertices in two buckets. The -1 elements in $B[0]$ were previously deleted.

requests we need to do an exclusive prefix sum of an array of request counts, so that each thread knows where to start writing. In our example this produces $[0, 1]$, and both threads write to a shared array, producing $[(3, 5), (1, 2), (4, 6)]$.

Next, consider building a request set starting from vertices 5 and 6, with distances 5 and 8, respectively. This gives a request set of $[(6, 7), (7, 9), (7, 9)]$. However, we don't want duplicates in our request sets. They occupy extra space, require synchronization when doing relaxations and insertion into buckets, and they could lead to multiple moves of the same vertex between buckets.

Avoiding duplicates requires the use of both sparse and dense representations, and a change in the way relaxation is handled, as shown in procedure `parRelax` in Algorithm 6.4. Instead of building a request set from all out-edges, we only add vertices that meet the condition $x < D[w]$, where $x = D[v] + c_{vw}$, in which case we also update $D[w] \leftarrow x$. The request is now simply the *id* of the vertex to be moved. The minimum of x and $D[w]$ is written atomically to $D[w]$, using procedure `atomicMin` (line 29), which in turn is implemented using the compare-and-swap (CAS) primitive [63]. The `atomicMin` procedure atomically updates $D[w]$ if it less than $D[v] + c_{vw}$ and returns **true**, or returns **false** otherwise. This ensures that $D[w]$ will contain the correct minimum distance if multiple threads update the same value. We avoid duplicates in the request set by using a CAS to update the sparse `visited` array (line 30). If the CAS fails, then that vertex is already in the subset and so doesn't need to be added to the dense array again. If it succeeds, then the vertex w is added to the request set *Req*.

In order to insert vertices into *Req* (line 31), each thread needs to know where in the array to begin writing. This can be implemented by each thread placing its requests in private arrays, then an exclusive prefix sum of the sizes of these arrays would indicate the starting point for each thread to write its requests. The `relax` function of Algorithm 6.3 is replaced by the `move` function that moves the vertex w from the request set according to the updated $D[w]$.

Bucket Data Structure

Since we know that there are no duplicate requests, there won't be any race conditions when moving vertices between buckets. The data structure illustrated in Figure 6.7 supports parallel insertions and deletions [47]. The buckets themselves are arrays that can be

Algorithm 6.4: Shared Memory Delta-Stepping Algorithm

Input: Graph with vertices V ($|V| = n$) and edges E with weights C, source vertex s.
Output: D: distance between s and all other vertices.

1: **parallel for** *vertex* $v \in V$ **do** //classify edges as light or heavy
2:    $H[v] \leftarrow \{e_{vw} \in E \mid c_{vw} > \Delta\}$
3:    $L[v] \leftarrow \{e_{vw} \in E \mid c_{vw} \leq \Delta\}$
4:    $D[v] \leftarrow \infty$
5:    $Bmap[v] \leftarrow -1$// map vertices to buckets
6: **end**
7: $D[s] \leftarrow 0$
8: move(s, $Bmap$)
9: **while** B *is not empty* **do**
10:    $R \leftarrow \emptyset$
11:    $visited \leftarrow \emptyset$
12:    **while** $B[i] \neq \emptyset$ **do**
13:       $Req \leftarrow$ parRelax($B[i]$, L)
14:       $R \leftarrow R \cup B[i]$
15:       $B[i] \leftarrow \emptyset$
16:       **parallel for** $w \in Req$ **do**
17:          move(w, $Bmap$)
18:       **end**
19:    **end**
20:    $Req \leftarrow$ parRelax(R, H)
21:    **parallel for** $w \in Req$ **do**
22:       move(w, $Bmap$)
23:    **end**
24:    $i \leftarrow i + 1$
25: **end**

// relax edges and create requests for bucket moves
26: **Procedure** parRelax($B[i]$, *kind*)
27:    **parallel for** *vertex* $v \in B[i]$ **do**
28:       **parallel for** *edge* $e_{vw} \in kind[v]$ **do**
          // atomically update $D[w]$
29:          **if** atomicMin($D[w]$, $D[v] + c_{vw}$) **then**
30:             **if** CAS($visited[w]$, *0*, *1*) **then**
31:                $Req \leftarrow Req \cup \{w\}$// parallel set insertion
32:             **end**
33:          **end**
34:       **end**
35:    **end**
36:    **return** Req
37: **end**

38: **Procedure** move(w, $Bmap$)
39:    **if** $Bmap[w] \neq -1$ **then**
40:       $B[Bmap[w]] \leftarrow B[Bmap[w]] \setminus \{w\}$
41:    **end**
42:    $B[\lfloor D[w]/\Delta \rfloor] \leftarrow B[\lfloor x/\Delta \rfloor] \cup \{w\}$
43:    $Bmap[w] \leftarrow \lfloor D[w]/\Delta \rfloor$
44: **end**

dynamically resized. There are also two mapping arrays: $Bmap$ to indicate in which bucket a vertex is stored and $Bloc$ indicates the position in the bucket array where the vertex is stored. Vertices that are not in any bucket are represented by -1 in the mapping arrays. Deletion of a vertex from a bucket is handled by setting its location in the bucket to -1. The use of $Bmap$ is shown in Algorithm 6.4. The $Bloc$ array is used when vertices are inserted into a bucket (line 42). Furthermore, a preprocessing sorting step can be added so that light and heavy edges from each vertex are stored contiguously. This means all edges of the same type can be more easily visited (line 28).

We can handle bucket moves in bulk by having each thread first write to temporary local arrays. Then the shared buckets can be expanded, a prefix sum used to compute the offset for each thread writing to each bucket, followed by parallel writes to the shared bucket. The same data structure can be used for the remember set R, which is used to remember vertices placed in the current bucket during relaxation of light edges so that heavy edges can be relaxed (line 20).

6.3.2 SIMD Bellman-Ford for GPU

Bellman-Ford is a good candidate for GPU implementation, since there is plenty of work for the massively parallel resources. The same `atomicMin` and `CAS` operations as above are supported on GPUs and can be used to provide atomic updates to the distance and vertex subset arrays. Making use of the CSR representation will allow us to easily strip off edges from the vertices in the active subset. Let's look at an example based on the graph represented in Figure 6.3.

Figure 6.8 traces through the SIMD Algorithm 6.5, starting from active vertices $F_1 =$

Algorithm 6.5: SIMD Bellman-Ford Algorithm

Input: Graph with n vertices and edges with nonnegative weights W, source vertex s. Vertices and edges stored in the CSR representation with offset array E_O and edge array E.

Output: D: distance between s and all other vertices.

1: $\{D[v] \leftarrow \infty : i \in [0, n)\}$
2: $\{D[s] \leftarrow 0\}$
3: $\{F_1[0] \leftarrow s\}$
4: **while** $F_1 \neq \emptyset$ **do**
5:     $\{F_N[i] \leftarrow E_O[F_1[i]] : i \in [0, |F_1|)\}$
6:     $\{S[i] \leftarrow E_O[F_1[i] + 1] - E_O[F_1[i]] : i \in [0, |F_1|)\}$
7:     $S_O \leftarrow \texttt{scan}(\text{exclusive, sum}, S)$
8:     $\{F_O[S_O[i] + j] \leftarrow E[F_N[i]] : i \in [0, |F_1|), j \in [0, S[i])\}$
9:     $\{W_O[S_O[i] + j] \leftarrow W[F_N[i]] : i \in [0, |F_1|), j \in [0, S[i])\}$
10:    $\{F_I[S_O[i] + j] \leftarrow F_1[i] : i \in [0, |F_1|), j \in [0, S[i])\}$
11:    $\{B \leftarrow \texttt{atomicMin}(D[F_O[i]], D[F_I[i]] + W_O[i])\ i \in [0, |F_O|)\}$
12:    $\{T[i] \leftarrow 0 : i \in [0, n)\}$
13:    $\{B \leftarrow \texttt{CAS}(T[F_O[i]], 0, 1) : i \in [0, |F_O|) \mid B[i] = 1\}$
14:    $B_S \leftarrow \texttt{scan}(\text{exclusive, sum}, B)$
15:    $\{F_2[B_S[i]] \leftarrow F_O[i] : i \in [0, |F_O|) \mid B[i] = 1\}$
16:    Swap references to F_1 and F_2
17: **end**

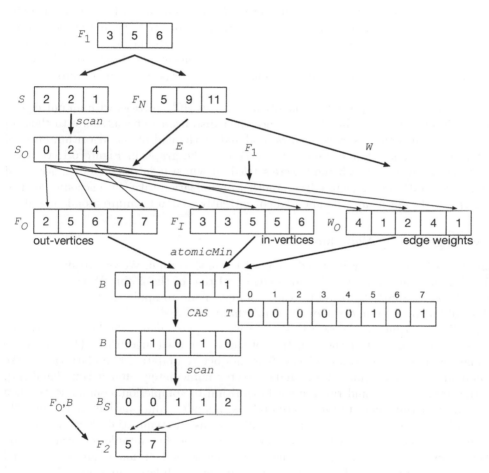

Figure 6.8: SIMD Bellman-Ford example, based on Figure 6.1.

$\{3, 5, 6\}$. It extracts the edges into two vertex arrays F_O, F_I and weight array W_O, performs the relaxations using `atomicMin`, removes duplicates using a `CAS`, and stores the new active vertices in F_2. The first step is to extract the edges from the vertices in F_1. This is done by extracting the offsets into the edge array E for each of the vertices into F_N and the number of out-edges into S. The target vertices are copied from the edge array and placed in F_O, where the starting index for each group is determined from S_O, a prefix sum of S. For example, the out-vertices corresponding to in-vertex 5 are written starting at index 2 of F_O. Similarly, the in-vertices for each edge are stored in F_I and the weights are stored in W_O. The relaxations using `atomicMin` update distance array D and return 1 for each vertex that has its distance updated. Here, vertex 7 has been successfully updated twice (one thread could have updated from ∞ to 10 and another from 10 to 9), but we don't want to insert it twice into the set of active vertices. To avoid such duplicates nonzero elements of B are updated using a `CAS` and an array T. Finally, a prefix sum of B provides the offsets into the new active vertex array F_2, and elements of F_O corresponding to nonzero elements of B are written to F_2.

Algorithm 6.5 can be improved by classifying vertices [10]. Vertices with an out-degree equal to 0 only need to be updated once, since their distance can't contribute to those of other vertices. Instead these vertices can be updated at the end of the algorithm once all other vertices are settled. For instance, in the case of the graph in Figure 6.1, vertex 7 could be updated once vertices 5 and 6 were settled, at which point its final distance could be calculated from its two incident edges. Another optimization takes into account vertices with in-degree of 1, which can't be updated concurrently so don't require atomic updates.

6.3.3 Message Passing Algorithm

Distributed SSSP algorithms rely on a decomposition of the graph. We've already discussed in Section 6.2 the characteristics of a good edge partition of the graph. What we need to do next is choose a SSSP algorithm for each compute node to apply to its subgraph, and choose how often to communicate boundary information.

The SSSP algorithm we choose to use at each compute node, whether sequential or multithreaded, does not affect the structure of the distributed algorithm. The choice we have to make is how long to run the local algorithm before updating boundary values. We could communicate the distance of a boundary vertex immediately after it is updated [11]. At the other extreme we could run the local algorithm to completion, communicate, then run the local algorithm again to propagate updated boundary information, and so on [39]. The first option will do the same computational work as the sequential algorithm, at the cost of frequent small messages between nodes. This could be suitable for execution on a parallel computer with a low latency network. The second option requires additional relaxations, but requires much fewer communication steps with larger messages. This could be attractive for execution on a more loosely coupled parallel platform with higher network latency. A good compromise is to let the local algorithm run for a fixed number of steps before communicating [49].

Let's assume that we start with the graph at one compute node, and a partition that assigns each edge to a compute node. Two examples are shown in Figures 6.9 and 6.10 for two partitions of the graph in Figure 6.1. Five arrays are constructed from the CSR representation of the graph and the edge partition for each subgraph. The edge and offset arrays are for the the CSR representation. The vertex map allows the vertex corresponding to each offset to be identified. This array is needed because the global numbering of vertices is preserved, and as a result the index of an element in the offset array doesn't correspond

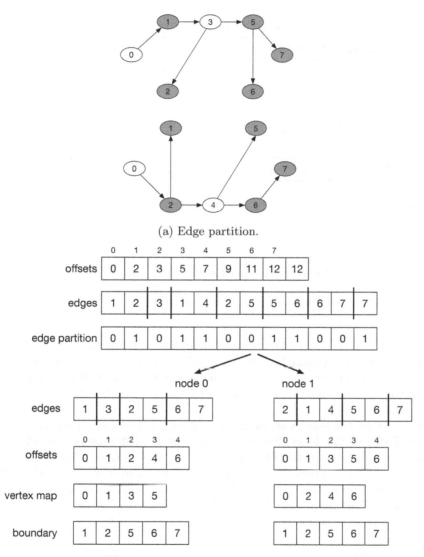

(a) Edge partition.

(b) Data structures for each subgraph.

Figure 6.9: First edge partition example. The CSR representation of Figure 6.3 is reproduced for convenience. Boundary vertices are indicated in gray.

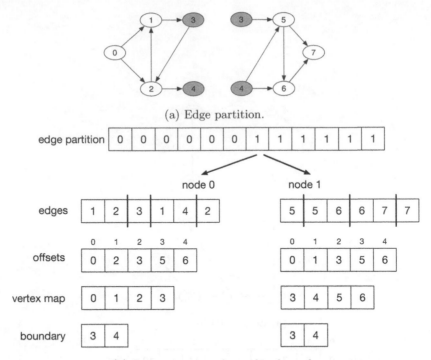

(a) Edge partition.

(b) Data structures for each subgraph.

Figure 6.10: Second edge partition example.

to the vertex *id*. We also need to identify the boundary vertices for each subgraph. An array of boundary vertices is assembled for each subgraph boundary, by searching for the vertices (out-vertices from the edge array and in-vertices from the vertex map array) that each subgraph has in common with other subgraphs. This also allows the neighbors of each node to be identified.

Once this accounting is taken care of and the arrays sent to each node, Algorithm 6.6 is quite straightforward. The local SSSP solver proceeds for t iterations. This algorithm is slightly modified to identify updates of boundary vertices. The value of t is chosen to find a compromise between communication frequency and number of relaxations. The distances of updated boundary vertices are exchanged between neighbors. Only distances that have been updated are sent to neighbors. This requires sending variable length messages, and identifying the length of each message received. If a received vertex distance is less than the local copy, then the vertex distance is updated and it is inserted into the list of active vertices of the local solver.

Termination of the algorithm occurs when all the vertex lists are empty. This is detected by performing an `allReduce` collective communication at the end of each iteration. This performs a Boolean **and** reduction on the local values and places the result on all nodes.

The largest drawback of this algorithm is that the amount of work for each compute node is not predicable in advance, which can result in considerable load imbalance. The graph partition significantly influences the load imbalance, and therefore should be carefully chosen. The partition has to find a compromise between load balance and communication. For instance, one could consider a random assignment of edges to compute nodes in order to

---

Algorithm 6.6: SSSP with Message Passing.

Input: Graph with n vertices and edges with nonnegative weights W, source vertex s. Edge partition of graph into p subgraphs.

Output: D: distance between s and all other vertices, distributed across p nodes.

1: **if** $id = 0$ **then**
2:     construct arrays for each subgraph from partition
3:     distribute arrays to compute nodes
4:     broadcast source vertex id to all nodes
5: **end**
6: **repeat**
7:     solve SSSP on subgraph for t iterations and mark boundary vertices that were updated
8:     send distances of updated boundary vertices to each neighbor
9:     receive updated distances from each neighbor
10:     insert received boundary vertices with improved distance in active vertex list
11:     **if** *active vertex list is empty* **then**
12:         localFinished ← true
13:     **else**
14:         localFinished ← false
15:     **end**
16:     allReduce(localFinished, *1*, and, finished)
17: **until** finished

---

balance the load. However, this would result in subgraphs with many connected components and would require frequent communication in order to avoid idle nodes.

6.4 CONCLUSION

We have seen three representative SSSP parallel algorithms. They have illustrated the trade-off between work, parallelism, and communication. The Delta-Stepping algorithm provides an approach to balancing work and parallelism, and forms the basis for many implementations. Load balancing is relatively straightforward for shared memory and SIMD algorithms, but the required atomic operations can negatively impact performance. Load balancing is much more difficult for distributed memory algorithms, and is very dependent on the characteristics of the graph and its partition.

Both the parallel loop and SIMD algorithms made use of operations on sets, represented as arrays. Race conditions were avoided using the compare-and-swap atomic operation. These algorithms provided a typical use for the prefix sum in computing offsets needed for parallel writes to a shared array. The message passing algorithm follows a common pattern in SPMD programs of iterative refinement, where tasks alternate between computing and communication and collectively synchronize to determine whether to proceed to the next iteration. Here there was a tradeoff between communicating less frequently and needing more iterations to finish.

The choice of graphs to use in experimental assessment of an algorithm is very important. Graphs of varying sizes, degree distributions, and diameters must be included. Various graph generators and graph libraries are available. In particular, the RMAT graph generator (`http://www.graph500.org/`) is particularly useful for creating scale-free graphs. Note that RMAT produces self-loops and multi-edges which need to be pruned to use for the SSSP.

Performance results are presented in traversed edges per second (TEPS), which is given by the number of edges of the graph divided by the run time. Crucially it's not the number of relaxations per second!

Graph algorithms such as SSSP have many practical applications, so it's not surprising that many graph processing frameworks have been created, such as Google's Pregel, and its open source alternative, Giraph (`http://giraph.apache.org/`). Shun and Blelloch's Ligra [63] is an attractive shared-memory framework. While MapReduce can be used for SSSP algorithms, it's easier to use the specialized frameworks and they offer much better performance.

6.5 FURTHER READING

There is a large body of literature on parallel SSSP algorithms, but most are elaborations of the techniques presented here. *Scalable Single Source Shortest Path Algorithms for Massively Parallel Systems*[11] has a nice discussion of the tradeoff in sequential algorithms between the number of phases and relaxations. The first chapter of Christian Sommer's PhD thesis, *Approximate Shortest Path and Distance Queries in Networks* (available at `http://www.sommer.jp/thesis.htm`) gives a very readable overview of the range of graphs that are encountered in practical shortest path problems. A good introduction to graph partitioning is given in Chapter 18 of the *Sourcebook of Parallel Computing* [21].

6.6 EXERCISES

6.1 Why is TEPS (traversed edges per second) not measured as relaxations per second?

6.2 Trace the sequential delta-stepping algorithm for Figure 6.1, for $\Delta = 3$. How does the number of relaxations (calls to `relax` procedure) compare with the execution with $\Delta = 5$ in Table 6.3?

6.3 Design an algorithm for the `writemin` procedure used in Algorithms 6.4 and 6.5, making use of the Compare and Swap function (Section 4.4.2).

6.4 Implement parallel set operations required by the Bellman-Ford and delta-stepping algorithms.

6.5 In this chapter we looked at a shared memory delta-stepping algorithm and a SIMD Bellman-Ford Algorithm. Design a shared memory Bellman-Ford algorithm and implement it for execution on a multiprocessor. Test with suitably large scale-free graphs, and determine the speedup compared to a sequential implementation of Dijkstra's algorithm.

6.6 Compare the bread-first search (BFS) algorithm with SSSP algorithms, explaining how the former provides a lower bound on execution time for the latter. Design and implement a simple shared memory BFS algorithm, and compare its runtime with the Bellman-Ford algorithm in Exercise 6.5 on the same graphs.

6.7 Implement the SIMD Bellman-Ford Algorithm 6.5 for GPU execution using CUDA or OpenCL. How does the performance depend on the degree distribution of the graphs?

6.8 Explore the sensitivity of distributed memory SSSP to the partition of the graph by implementing Algorithm 6.6 with the modified Bellman-Ford algorithm and using MPI for communication. The graph can be partitioned by dividing the edge array into p concurrent chunks (as was done in Figure 6.10). Create multiple partitions of the same graph by permuting the vertex numbering. How do the communication and computation times depend on the number of boundary vertices?

CHAPTER 7

The Eikonal Equation

Wave propagation is everywhere in nature and is also a very useful problem solving tool. Take for instance an indirect way to measure the distance between two points in two dimensions by propagating a front from the source point. The front will move in concentric circles until the destination point is reached, as illustrated in Figure 7.1. If the front moves with unit speed, the time of arrival gives the distance between the two points. We can also plot the shortest path between the two points by tracing a line moving back from the destination to the source point while keeping perpendicular to the contours.

This might seem like an odd way to find the distance between two points in a plane. But what if the speed wasn't equal to one everywhere, but instead varied according to some function? The front propagation technique can easily solve this problem in exactly the same way. In Figure 7.2 the speed function is 0.5 in the top part of the image and 1 elsewhere. The boundary between the two values can be seen in the faint horizontal line, and the darker color in the top portion indicates that the front takes longer to propagate there. We can also put obstacles, by assigning a speed of 0 to a region, through which nothing can pass. In Figure 7.3 the obstacle increases the path to the upper half of the domain.

Tracking moving fronts turns out to be very useful in solving many real-world problems, such as seismic imaging, boundary detection in image processing, robotic navigation, and modeling the movement of crowds. The last problem has been applied to simulating crowd movement in computer games [62].

Several methods exist to track moving fronts, but *level set methods* have been the most successful [60]. This chapter refers to a special case where the time dependent problem is turned into a stationary problem. This can be done when the front always moves forward or backward, so the front only passes by a given point once, and we can view the solution as a static contour plot, as in Figures 7.1–7.3. The plot of the stationary front is obtained by solving the *Eikonal equation*:

$$|\nabla U(\mathbf{x})| = \frac{1}{F(\mathbf{x})}, \ \mathbf{x} \in \Re^n, \tag{7.1}$$

where ∇ denotes the gradient, $| \, |$ the Euclidean norm, $F(\mathbf{x})$ is a positive speed function, and the boundary condition is given by the initial position of the front: $U(\mathbf{x}) = 0, \mathbf{x} \in \Gamma \subset \Re^n$. The solution $U(\mathbf{x})$ gives the arrival time at position $\mathbf{x}$. In this chapter we'll stick to two dimensions, and with single point boundaries.

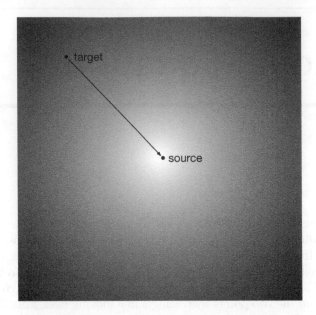

Figure 7.1: Distance between two points using a propagating front. Shade of gray indicates arrival time, with shade getting darker as time increases from 0 (white). Shortest path is found by tracing back to the source, keeping perpendicular to the front.

Figure 7.2: Arrival times of front starting from the center point, in a domain with 2 speeds: 0.5 in top 40% and 1 elsewhere. Darker shades in the top half indicate that arrival times are greater.

Figure 7.3: Arrival times of front starting from the center point, in a domain with an obstacle (white rectangle, speed = 0), showing how the front moves around the obstacle.

7.1 NUMERICAL SOLUTION

Partial differential equations are solved by discretizing the domain, which can be done by tiling it with polygons such as rectangles and triangles. We'll use a discretization of a square domain into $(n_i - 1) \times (n_j - 1)$ squares of length h, and solve the Eikonal equation at the corners of the squares. The solution $u_{i,j}$ is initialized to ∞, which in practice can be a value larger than the upper bound of U, except $u_{i,j}$ is initialized to zero at the points that are on the boundary (source of the front). We'll also add another layer of points around the edge of the domain, with $u_{0,0..n_j+1} = u_{n_i+1,0..n_j+1} = u_{0..n_i+1,0} = u_{0..n_i+1,n_j+1} = \infty$, which ensures that the solution flows outward at the edges (see below). For example, a $n_i = 7 \times n_j = 7$ domain with the boundary in the center point, has u initialized to 100 (our stand-in for ∞) in a 9×9 grid in Table 7.1.

Table 7.1: Initial grid for solution to Eikonal equation, including extra layer at edges.

100	100	100	100	100	100	100	100	100
100	100	100	100	100	100	100	100	100
100	100	100	100	100	100	100	100	100
100	100	100	100	100	100	100	100	100
100	100	100	100	**0**	100	100	100	100
100	100	100	100	100	100	100	100	100
100	100	100	100	100	100	100	100	100
100	100	100	100	100	100	100	100	100
100	100	100	100	100	100	100	100	100

The Eikonal equation is solved numerically by replacing the continuous gradient ∇ by a finite difference approximation at the points on our discrete grid, resulting in an approximation u of the exact solution U. Let's start by looking at the derivative $\partial U / \partial y$, which can be approximated at (i, j) by $(u_{i,j} - u_{i',j})/h$, where i and i' are neighboring points separated by h. How to choose i'? We can observe that solutions of the Eikonal equation are biased in a given direction, as the front propagates either forward or backward. To reflect this bias we select the neighboring point i' that is in the *upwind* direction, which gives:

$$\frac{\partial U}{\partial x} \approx \frac{u_{i,j} - u^{imin}}{h}, \ u^{imin} = \min(u_{i-1,j}, u_{i+1,j})$$

In our case, upwind means pointing back to the boundary (source of the front), which means pointing to smaller values of u. Doing the same thing in the x direction and using the definition of the Euclidean norm gives a discretized version of the Eikonal equation [77]:

$$\left((u_{i,j} - u^{imin})^+\right)^2 + \left((u_{i,j} - u^{jmin})^+\right)^2 = \frac{h^2}{f_{i,j}^2}, \tag{7.2}$$

where $f_{i,j}$ is the speed function F evaluated at (i, j), and

$$(x)^+ = \begin{cases} x, & x > 0, \\ 0, & x \le 0. \end{cases}$$

The $(\ )^+$ operator is needed because $u_{i,j} - u^{xmin}$ and $u_{i,j} - u^{ymin}$ can't be negative, since the solution increases in the downwind direction. The reason for the extra layer of points with $U = \infty$ should now be clear, as it ensures that the solutions flow toward the edges of the domain.

We now have a set of quadratic equations to solve. The solution to Equation 7.2 at each point is [77]:

$$u_{i,j}^{new} = \begin{cases} \min(u^{imin}, u^{jmin}) + h/f_{i,j}, & |u^{imin} - u^{jmin}| \ge h/f_{i,j}, \\[2ex] \frac{u^{imin} + u^{jmin} + \sqrt{2h^2/f_{i,j}^2 - (u^{imin} - u^{jmin})^2}}{2}, & |u^{imin} - u^{jmin}| < h/f_{i,j}. \end{cases} \tag{7.3}$$

The second case of Equation 7.3 is the usual formula for the solution to the quadratic equation. We have excluded the other solution, $\left(-b - \sqrt{b^2 - 4ac}\right)/2a$, since it is inconsistent with the conditions $u_{i,j} \ge u^{imin}$, $u_{i,j} \ge u^{jmin}$, at least one of which must be true according the $()^+$ operator in Equation 7.2 . You can verify the solution for the first case by plugging it into Equation 7.2.

Algorithms for solving the Eikonal equation over the whole domain differ in the order in which the points are visited, and each has different implications for parallelization.

7.1.1 Fast Sweeping Method

Let's visit the interior points on the grid in Table 7.1 left to right, a row at a time, starting from the bottom row, replacing $u_{i,j}$ by $u_{i,j}^{new}$ at each point if $u_{i,j}^{new} < u_{i,j}$, using $F = 1$ everywhere and $h = 1$. This results in the solution in Table 7.2, where we have omitted the extra layer around the domain.

Even though we have visited each point of the domain, only the values in the upper right quadrant (intersection of first four rows and last four columns) are correct. This is because the direction of this sweep points to the upper right (see Figure 7.4a) and the front has propagated from the center point along this direction.

This type of sweep is reminiscent of the Gauss-Seidel method for solving systems of

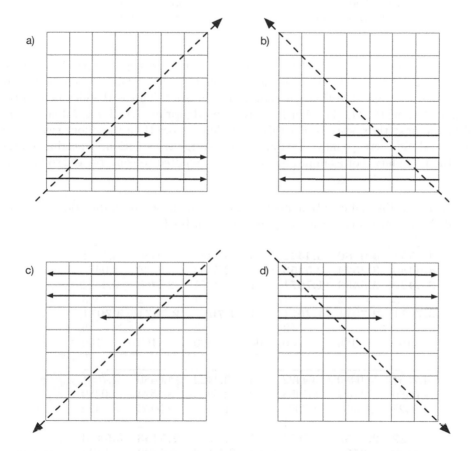

Figure 7.4: Four sweeps in the Fast Sweeping Method in two dimensions. In each case a partial traversal of the grid is shown with solid arrows and the overall direction is shown with a dashed arrow.

Table 7.2: Solution to Eikonal equation for grid in Table 7.1, with outer layer of points not shown, after first sweep in direction given by Figure 7.4a. $F = 1$ everywhere and $h = 1$. New values are shown in bold.

100	100	4	3	**3.4422**	**4.0480**	**4.7551**
100	100	3	2	**2.5453**	**3.2524**	**4.0480**
100	100	2	1	**1.7071**	**2.5453**	**3.4422**
100	100	1	0	1	2	3
100	100	100	1	2	3	4
100	100	100	100	100	100	100
100	100	100	100	100	100	100

linear equations. As in that method, here we need to perform sweeps iteratively until the solution converges. We can be clever about the additional sweeps by having them propagate fronts in the other diagonal directions, as shown in Figure 7.4. This is the idea behind the Fast Sweeping Method (FSM) [77], which is given in Algorithm 7.1. Table 7.3 shows the solution after the next three sweeps. For this example no more than these four sweeps are needed. In general more sweeps may be required to converge the solution, depending on the geometry of the boundary. For example, 5 sweeps were required to converge the solution in Figure 7.3.

Table 7.3: Solution to Eikonal equation after second to fourth sweeps in directions given by Figure 7.4b–d. New values after each sweep are shown in bold.

4.7551	**4.0480**	**3.4422**	3	3.4422	4.0480	4.7551
4.0480	**3.2524**	**2.5453**	2	2.5453	3.2524	4.0480
3.4422	**2.5453**	**1.7071**	1	1.7071	2.5453	3.4422
3	**2**	1	0	1	2	3
3.7071	**2.7071**	**1.7071**	1	1.7071	**2.7071**	**3.7071**
5	**4**	**3**	2	**3**	**4**	**5**
100	100	100	100	100	100	100

4.7551	4.0480	3.4422	3	3.4422	4.0480	4.7551
4.0480	3.2524	2.5453	2	2.5453	3.2524	4.0480
3.4422	2.5453	1.7071	1	1.7071	2.5453	3.4422
3	2	1	0	1	2	3
3.4422	**2.5453**	1.7071	1	1.7071	**2.5453**	**3.5453**
4.0480	**3.2524**	**2.5453**	2	**2.5453**	**3.4422**	**4.4422**
4.7551	**4.0480**	**3.4422**	3	**3.5453**	**4.4422**	**5.4422**

4.7551	4.0480	3.4422	3	3.4422	4.0480	4.7551
4.0480	3.2524	2.5453	2	2.5453	3.2524	4.0480
3.4422	2.5453	1.7071	1	1.7071	2.5453	3.4422
3	2	1	0	1	2	3
3.4422	2.5453	1.7071	1	1.7071	2.5453	**3.4422**
4.0480	3.2524	2.5453	2	2.5453	**3.2524**	**4.0480**
4.7551	4.0480	3.4422	3	**3.4422**	**4.0480**	**4.7551**

The complexity of the Fast Sweeping Method is $O(n^2)$ for a $n \times n$ grid, as the number of iterations required to converge is a constant that depends on the geometry of the

---

Algorithm 7.1: Two dimensional Fast Sweeping Method

Input: Grid spacing h, $(n_i + 2) \times (n_j + 2)$ speed function matrix F,
$(n_i + 2) \times (n_j + 2)$ solution matrix U initialized to large value everywhere
except boundary (source of front), where it is initialized to 0.

Output: Solution matrix U

while *not converged* **do**
    sweep(U, n_i, 1, 1, n_j, F, h) // Northeast
    sweep(U, n_i, 1, n_j, 1, F, h) // Northwest
    sweep(U, 1, n_i, n_j, 1, F, h) // Southwest
    sweep(U, 1, n_i, 1, n_j, F, h) // Southeast
end

Procedure sweep(U, i_a, i_b, j_a, j_b, F, h)
    if $i_b < i_a$ **then** $step_i = -1$ **else** $step_i = 1$
    if $j_b < j_a$ **then** $step_j = -1$ **else** $step_j = 1$
    for $i \leftarrow i_a$ to i_b **step** $step_i$ **do**
        for $j \leftarrow j_a$ to j_b **step** $step_j$ **do**
            $u_{\text{new}} \leftarrow$ solveQuadratic(U, i, j, F, h)
            if $u_{new} < U[i,j]$ **then** $U[i,j] \leftarrow u_{\text{new}}$
        end
    end
end

Procedure solveQuadratic(U, i, j, F, h)
    // Don't update boundary points
    if $U[i,j] \leftarrow 0$ **then**
        return $U[i,j]$
    end
    $a \leftarrow$ min($U[i-1,j]$, $U[i+1,j]$)
    $b \leftarrow$ min($U[i,j-1]$, $U[i,j+1]$)
    if $|a - b| \geq h/F[i,j]$ **then**
        return min(a,b) $+ h/F[i,j]$
    else
        return $\left(a + b + \sqrt{2 * (h/F[i,j])^2 - (a-b)^2} \right) / 2$
    end
end

---

boundary and not on the problem size. It is very easy to implement, as you can see from Algorithm 7.1. In practice, the total number of arithmetic operations will depend on which case of Equation 7.3 is evaluated at each point. The number of evaluations of this equation can be reduced by skipping points that have all neighbors with $U = \infty$. Also, in cases where the front only needs to be propagated for a limited time L, points only need to be updated when $\min(u^{imin}, u^{jmin}) < L$.

7.1.2 Fast Marching Method

The solution to the Eikonal equation gives the continuous shortest path from the boundary to all other points in the domain. So it's not surprising that an algorithm closely related to Dijkstra's single source shortest path algorithm (see Chapter 6) produces an optimal solution (in terms of number of points visited) to the discretized Eikonal Equation 7.2. The Fast Marching Method (FMM) [59] visits each point once, starting from the boundary and proceeding along the shortest path, using Equation 7.3 to compute the cost of reaching the neighbors of the current point.

The method starts by solving Equation 7.3 for all points adjacent to the boundary. Beginning again with our 7×7 example, with $F = 1$ and $h = 1$, this gives Table 7.4. Each point is labeled as being either KNOWN, BAND, or FAR. All boundary points are KNOWN (the center point in Table 7.4) and neighbors of KNOWN points are in the BAND. All other points are FAR. As shown in Algorithm 7.2 [75], the next step is to select the BAND point with the smallest value, change its label to KNOWN, and compute the values of its neighbors. When solving Equation 7.3, only those neighbors that are KNOWN are included, as shown in procedure `selectMin` in Algorithm 7.3. The values of the BAND points are stored in an indexed priority queue, which allows selection of the point with the minimum value at each iteration and updating of points already in the queue.

Table 7.4: First step of Fast Marching method, $F = 1$ everywhere and $h = 1$.

100	100	100	100	100	100	100
100	100	100	100	100	100	100
100	100	100	1	100	100	100
100	100	1	0	1	100	100
100	100	100	1	100	100	100
100	100	100	100	100	100	100
100	100	100	100	100	100	100

Table 7.5 shows the evolution of the matrix in Table 7.4 after each of the first 2 iterations, and after iteration 10 and 11. After iteration 10 the top value in the queue has index $(4, 2)$ and value 2, so this point becomes KNOWN in iteration 11. After 48 iterations the final solution is reached. Note that the number of iterations reflects the number of points in the domain, excluding the boundary ($7 \times 7 - 1$).

The complexity of the Fast Marching Method is $O(n^2 \log n^2)$, since the reordering of a priority queue of length m implemented as a binary heap takes $O(\log m)$ and the size of the queue is $O(n^2)$. This method has a higher complexity than the Fast Sweeping Method, because of the cost of maintaining a priority queue. In practice the marching method is often faster than the sweeping method, particularly when the latter takes a relatively large number of iterations to converge. The marching method requires fewer evaluations of the quadratic solver and the size of the queue is usually much less than n^2 [75].

Algorithm 7.2: Fast Marching Method

Input: Grid spacing h, desired width L of solution, $(n_i + 2) \times (n_j + 2)$ speed function matrix F, $(n_i + 2) \times (n_j + 2)$ solution matrix U initialized to large value everywhere except boundary (source of front), where it is initialized to 0.

Output: Solution matrix U

Data: $(n_i + 2) \times (n_j + 2)$ matrix G, with possible values KNOWN, BAND, FAR. Initialized to FAR everywhere except boundary, where it is initialized to KNOWN

Data: Indexed min priority queue iQmin

foreach *element (i, j) of G such that $G[i, j] =$ KNOWN* **do**
  updateNeighbors(iQmin, U, G, i, j, F, h)
end
while iQmin *is not empty* **do**
  $(l, m) \leftarrow$ iQmin.minIndex() // retrieve index of element at front of queue
  if $U[l, m] > L$ **then** break
  $G[l, m] \leftarrow$ KNOWN
  iQmin.delMin() // remove element at front of queue
  updateNeighbors(iQmin, U, G, i, j, F, h)
end

Procedure updateNeighbors(iQmin, U, G, i, j, F, h)
  for $(l, m) \leftarrow (i+1, j)$, $(i-1, j)$, $(i, j+1)$, $(i, j-1)$ **do**
    if $G[l, m] =$ KNOWN $\vee$ (l, m) *outside domain* **then** continue
    $u_{\text{temp}} \leftarrow$ solveQuadratic(G, U, l, m, F, h)
    if $u_{temp} < U[l, m]$ **then**
      $U[l, m] \leftarrow u_{\text{temp}}$
      $G[l, m] \leftarrow$ BAND
      if iQmin.contains((l, m)) **then**
        iQmin.change($(l, m), u_{temp}$)// update element in queue
      else
        iQmin.insert($(l, m), u_{temp}$)// insert element in queue
      end
    end
  end
end

Algorithm 7.3: Quadratic solver for Fast Marching Method

Procedure solveQuadratic(G, U, i, j, F, h)

    $a \leftarrow$ selectMin(G, U, $i+1$, j, $i-1$, j)

    $b \leftarrow$ selectMin(G, U, i, $j+1$, i, $j-1$)

    if $a = -1$ **then**

        return $b + h/F[i,j]$

    else if $b = -1$ **then**

        return $a + h/F[i,j]$

    else if $|a - b| \geq h/F[i,j]$ **then**

        return $\min(a,b) + h/F[i,j]$

    else

        return $\left(a + b + \sqrt{2 * (h/F[i,j])^2 - (a-b)^2} \right) / 2$

    end

end

Procedure selectMin(G, U, l, m, p, q)

    $x \leftarrow -1$

    if $G[l,m] = $ known $\wedge$ $G[p,q] = $ known **then**

        $x \leftarrow \min(U[l,m],\, U[p,q])$

    else if $G[l,m] \neq$ known $\wedge$ $G[p,q] = $ known **then**

        $x \leftarrow U[p,q]$

    else if $G[l,m] = $ known $\wedge$ $G[p,q] \neq$ known **then**

        $x \leftarrow U[l,m]$

    end

    return x

end

Table 7.5: Fast Marching Method solution to Eikonal equation after iterations 1, 2, 10, 11. Bold values indicate BAND points. Non-bold values other than 100 indicate KNOWN values.

100	100	100	100	100	100	100
100	100	100	100	100	100	100
100	100	100	**1**	100	100	100
100	100	**1**	0	**1**	100	100
100	100	**2**	1	**2**	100	100
100	100	100	**2**	100	100	100
100	100	100	100	100	100	100

100	100	100	100	100	100	100
100	100	100	100	100	100	100
100	100	**2**	1	100	100	100
100	**2**	1	0	**1**	100	100
100	100	**1.7071**	1	**2**	100	100
100	100	100	**2**	100	100	100
100	100	100	100	100	100	100

100	100	100	100	100	100	100
100	100	**2.7071**	2	**2.7071**	100	100
100	**2.7071**	1.7071	1	1.7071	**2.5453**	100
100	**2**	1	0	1	2	**3**
100	**2.7071**	1.7071	1	1.7071	**2.5453**	100
100	100	**2.5453**	2	**2.5453**	100	100
100	100	100	**3**	100	100	100

100	100	100	100	100	100	100
100	100	**2.7071**	2	**2.7071**	100	100
100	**2.5453**	1.7071	1	1.7071	**2.5453**	100
3	2	1	0	1	2	**3**
100	**2.5453**	1.7071	1	1.7071	**2.5453**	100
100	100	**2.5453**	2	**2.5453**	100	100
100	100	100	**3**	100	100	100

7.2 PARALLEL DESIGN EXPLORATION

Parallel algorithm design begins with a thorough understanding of the problem. A good understanding of existing sequential algorithms is necessary, but is not sufficient. It's also important to know the characteristics of the data that is used in practice. In the case of the Eikonal equation this includes the geometry of the boundary, the speed functions F that are used, and the size of the grids. The relative merits of different sequential and parallel algorithms depends on these characteristics. We also need to know how the program is used and what the goals of parallelization are. For instance, a parallel implementation may not be required if there are a large number of independent solutions of medium-sized grids, as independent sequential programs can be executed in parallel on a cluster. We could be dealing with a real-time problem, where strong scaling is needed to achieve solution by a deadline, or with offline processing of very large grids, where good weak scaling is desired.

The first design step is decomposition. Here the tasks can easily be identified as each updating a single element of the grid. This produces plenty of tasks that grow with the grid size. The dependencies between tasks seemingly prevent any parallel execution, however. Each task depends on its four immediate neighbors in the i and j directions. In each FSM sweep the element updates aren't independent. The sweeps in the four directions are additive, as can be seen in Table 7.3. We aren't any better off with FMM, as it proceeds by sequentially processing elements from the priority queue. This means that we need to examine how to modify the sequential algorithms to expose independent tasks.

7.2.1 Parallel Fast Sweeping Methods

There are two ways we can create independent tasks: relax the dependencies and pay the price with more iterations, or change the order in which we visit points.

Relaxing Dependencies

Relaxing dependencies is sometimes used in iterative numerical methods to increase the number of independent tasks. An easy way we can relax dependencies in our case is to perform the four FSM sweeps in parallel. This produces four values for each element, from which we select the minimum: $u_{\text{new}} = \min(u_{i,j}^a, u_{i,j}^b, u_{i,j}^c, u_{i,j}^d)$ [78]. Applying this method to our 7×7 example from above gives the same results for the first sweep (Table 7.2), but the remaining sweeps are shown in Table 7.6. The final result is the same as in Table 7.3.

Table 7.6: Solution to Eikonal equation after parallel execution to of second to fourth sweeps in directions given by Figure 7.4b-d. New values after each step are shown in bold. Final result is given by minimum value from each of four sweeps.

4.7551	**4.0480**	**3.4422**	3	4	100	100
4.0480	**3.2524**	**2.5453**	2	3	100	100
3.4422	**2.5453**	**1.7071**	1	2	100	100
3	2	1	0	1	100	100
4	3	2	1	100	100	100
100	100	100	100	100	100	100
100	100	100	100	100	100	100

100	100	100	100	100	100	100
100	100	100	100	100	100	100
4	3	2	1	100	100	100
3	2	1	0	1	100	100
3.4422	**2.5453**	**1.7071**	1	2	100	100
4.0480	**3.2524**	**2.5453**	2	3	100	100
4.7551	**4.0480**	**3.4422**	3	4	100	100

100	100	100	100	100	100	100
100	100	100	100	100	100	100
100	100	100	1	2	3	4
100	100	1	0	1	2	3
100	100	2	1	**1.7071**	**2.5453**	**3.4422**
100	100	3	2	**2.5453**	**3.2524**	**4.0480**
100	100	4	3	**3.4422**	**4.0480**	**4.7551**

In this case, taking the sweeps independently did not require any additional sweeps to converge, but it does require an extra traversal of the matrices to find the minimum values.

In general this method can take more sweeps to converge than the sequential algorithm, but the extra work is independent of the grid size [78]. It also has a larger memory footprint than sequential FSM, as it requires storing four copies of the grid.

The degree of concurrency is only 4 (or 8 for 3D FSM), so this method is not scalable. To go further requires breaking the dependencies during individual sweeps. This can be done by using a 2D data decomposition, where each task updates a rectangular portion of the grid [78]. Each task depends on neighboring tasks in the decomposition, and uses ghost cells to store elements needed from neighbors. This is very similar to the Game of Life decomposition we saw in Chapter 4, except here elements only depend on up to 4 neighbors instead of 8. Each task updates its grid elements independently over the four sweeps, then exchanges inter-subdomain border elements with its neighbors. It's also possible to exchange elements after every sweep, but this may introduce too much communication.

Table 7.7 shows an example where the decomposition requires an extra set of sweeps to converge. After the first set of sweeps, only the lower right subdomain has correct values, since it contained the boundary element. The top left subdomain has not changed at all, since it is not adjacent to any of the known initial values.

Table 7.7: Solution to Eikonal equation after parallel execution of Fast Sweeping Method in four subdomains. New values after second set of sweeps shown in bold.

100	100	100	3	3.5453	4.2524	5.0480
100	100	100	2	2.7071	3.5453	4.4422
100	100	100	1	2	3	4
3	2	1	0	1	2	3
3.5453	2.7071	2	1	1.7071	2.5453	3.4422
4.2524	3.5453	3	2	2.5453	3.2524	4.0480
5.0480	4.4422	4	3	3.4422	4.0480	4.7551

4.7551	**4.0480**	**3.4422**	3	3.4422	4.0480	4.7551
4.0480	**3.2524**	**2.5453**	2	2.5453	3.2524	4.0480
3.4422	**2.5453**	**1.7071**	1	1.7071	2.5453	3.4422
3	2	1	0	1	2	3
3.4422	**2.5453**	**1.7071**	1	1.7071	2.5453	3.4422
4.0480	**3.2524**	**2.5453**	2	2.5453	3.2524	4.0480
4.7551	**4.0480**	**3.4422**	3	3.4422	4.0480	4.7551

This approach not only requires additional sweeps to converge, but also does not scale well. As the number of subdomains increases, more will have no work to do, as with the top left subdomain of Table 7.7. Also, fronts are propagated for shorter distances as the size of subdomains decreases, requiring even more iterations. At the limit where a subdomain contains one element, $O(n^2)$ iterations are required!

Reordering Sweeps

Increasing the number of independent tasks by relaxing dependencies does not look very promising for parallelizing the Fast Sweeping Method. Recall that each sweep moves in one of four diagonal directions by sweeping along each row. As illustrated in Figure 7.5a the update of each element depends on the updated value in the previous element in the row. If we instead sweep directly along each diagonal then we can independently update

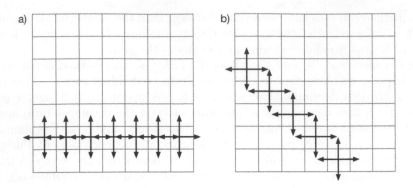

Figure 7.5: Two sweeping orders for Fast Sweeping Method: a) along each row b) along each diagonal. The second ordering permits independent updates.

7	8	9	10	11	12	13
6	7	8	9	10	11	12
5	6	7	8	9	10	11
4	5	6	7	8	9	10
3	4	5	6	7	8	9
2	3	4	5	6	7	8
1	2	3	4	5	6	7

Figure 7.6: Traversal order for parallel Fast Sweeping Method along each diagonal, corresponding to the first sweeping direction.

each element on a diagonal (Figure 7.5b). Thus we can perform each of the four sweeps by traversing diagonals of the grid in one of the four directions [18].

Consider sweeping in the first diagonal direction (Figure 7.4a), but with elements visited along each diagonal in parallel. The order of the traversal of the grid is illustrated in Figure 7.6. Notice that a cell is updated only after its left and bottom neighbors have been updated, as in the original order of Figure 7.4a. Therefore this ordering gives the same result as the sequential FSM, and no additional sweeps are needed, as they were with the previous parallel decompositions.

Analysis

The two dependence relaxing decompositions resulted in a potential increase in the number of iterations. The first one has $O(n^2)$ work and $O(n^2)$ depth, so it has only $O(1)$ parallelism. It can achieve some speedup of up to 4, although the speedup will be reduced by the extra work required to find the minimum and by any extra iterations needed. It's important to point out that our example has a trivial speed function $F = 1$. Practical problems will likely have more complex speed functions, so the overhead of finding the minimum may not be significant compared to the cost of computing F. The second decomposition results in an unknown increase $g(p)$ in the number of iterations as a function of the number p of subdomains. The author of this method suggests that $g(p)$ can be made close to 1 by having the ordering of sweeps vary between iterations [78]. This function will also depend on the geometry of the boundary. The work of this decomposition is $O(n^2g(p))$ and the depth is $O(n^2g(p)/p)$. This means that while the parallelism is of $O(p)$ it is not work-efficient if $g(p)$ is greater than $O(1)$.

The work of the third decomposition is exactly the same as the sequential algorithm, $O(n^2)$. The depth is given by the number of diagonals in a grid, which is $O(n)$, which means the parallelism is $O(n)$.

7.2.2 Parallel Fast Marching Methods

Fast Iterative Method

The Fast Marching Method is inherently sequential, so some relaxation of dependencies is needed to produce independent tasks. The approach of the Fast Iterative Method (FIM) [43] is to modify the FMM to allow updating of points in the wavefront (the BAND elements) in parallel. Coordinates of the band elements are stored in a simple list instead of a priority queue. New values of U are stored in a copy V so that updates can be done independently. Constrast this with the FSM, where updates of a single matrix could only take place independently along diagonals. Just as the update scheme in the FSM is the same as in the Gauss-Seidel method for linear equations, the FIM updates follow the same pattern as the Jacobi method for linear equations. Because we are not following the strictly increasing order of the priority queue the values of $u_{i,j}$ may need to be updated several times, thus making this an iterative method. Elements are kept in the active band list until their updated values are unchanged from the previous iteration, at which point they are removed from the band and their neighbors can be considered for addition to the list.

The FIM in Algorithm 7.4 starts by initializing the list L_1 of band elements with neighbors of boundary elements (lines 1–3). These elements are then updated, storing results in a copy V of the matrix. For our ongoing example this gives the same result as the FMM in Table 7.4. Elements in the band list are updated repeatedly. After each iteration, the neighbors of those elements that have converged values (and aren't already in the band list) are updated and the unconverged elements are added to a list (L_2) for the next iteration; if the update of a neighbor decreases its value it is added to the list. A value is considered to be converged if it changes by less that a small amount ϵ from one iteration to the next. References to U and V and to L_1 and L_2 are swapped at the end of each iteration.

Table 7.8 shows the matrix after the second and third iterations of the example begun in Table 7.4. One can see that band values move outward radially. The FIM requires each element to be updated at least twice, to check convergence, although in this example the first update produces the correct values.

An important detail of the FIM not revealed in our simple example is that values that are removed from the band list may be added again if a neighbor converges its value (line 10), that is, they are not KNOWN in the sense of the FMM, but are only provisional. To illustrate this we need to look at another example, such as the one with two boundary points shown in Table 7.9 after the first iteration. We can see in the remaining iterations in Table 7.10 that some points need to be re-added to the list, such as elements $(2,5)$ and $(6,2)$, which were re-added to the list when their values changed from 3 to 2.9974. Notice also that some values in the band change several times, such as the element $(1,6)$ which changes from 4.3709 to 4.3378 to 4.337. The complexity introduced by having 2 boundary points results in two additional iterations compared to the single boundary point case.

We can easily see that FIM involves doing duplicate work. Pairs of converged elements

---

Algorithm 7.4: Fast Iterative Method.

Input: Grid spacing h, $(n_i + 2) \times (n_j + 2)$ speed function matrix F, $(n_i + 2) \times (n_j + 2)$ solution matrix U initialized to large value everywhere except boundary (source of front), where it is initialized to 0.

Output: Solution matrix U

Data: Copy V of solution matrix U, initialized to same values.

Data: Two lists L_1 and L_2, containing elements in active band.

1: **foreach** *element (i, j) of U which is a neighbor of a boundary element* **do**
2:   add (i, j) to L_1
3: **end**
4: **while** *L_1 is not empty* **do**
5:   **foreach** *element (i, j) in L_1* **do**
6:    $u_{new} \leftarrow$ `solveQuadratic(`U`, `i`, `j`, `F`, `h`)` // same function as Alg. 7.1
7:    $V[i, j] \leftarrow u_{new}$
8:    **if** $|u_{new} - U[i, j]| < \epsilon$ **then**
9:     **for** $(l, m) \leftarrow (i+1, j)$, $(i-1, j)$, $(i, j+1)$, $(i, j-1)$ **do**
10:      **if** (l, m) *not in L_1* **then**
11:       $u_{new} \leftarrow$ `solveQuadratic(`U`, `l`, `m`, `F`, `h`)`
12:       **if** $u_{new} < U[l, m]$ **then**
13:        $V[l, m] \leftarrow u_{new}$
14:        add (l, m) to L_2
15:       **end**
16:      **end**
17:     **end**
18:    **else**
19:     add (i, j) to L_2 // keep in list for next iteration if not converged
20:    **end**
21:   **end**
22:   remove duplicates from L_2
23:   swap references to L_1 and L_2
24:   swap references to U and V
25: **end**

Table 7.8: Solution to Eikonal equation after second and third iterations of Fast Iterative Method, after same initialization as FMM shown in Table 7.4. Values in active band shown in bold. Three more iterations are required to complete solution over whole domain.

100	100	100	100	100	100	100
100	100	100	**2**	100	100	100
100	100	**1.7071**	1	**1.7071**	100	100
100	**2**	1	0	1	**2**	100
100	100	**1.7071**	1	**1.7071**	100	100
100	100	100	**2**	100	100	100
100	100	100	100	100	100	100

100	100	100	**3**	100	100	100
100	100	**2.5453**	2	**2.5453**	100	100
100	**2.5453**	1.7071	1	1.7071	**2.5453**	100
3	2	1	0	1	2	**3**
100	**2.5453**	1.7071	1	1.7071	**2.5453**	100
100	100	**2.5453**	2	**2.5453**	100	100
100	100	100	**3**	100	100	100

Table 7.9: First iteration of FIM of domain with 2 boundary points.

100	**1**	100	100	100	100	100
1	**0**	**1**	100	100	100	100
100	**1**	100	100	100	100	100
100	100	100	100	100	100	100
100	100	100	100	**1**	100	100
100	100	100	**1**	**0**	**1**	100
100	100	100	100	**1**	100	100

often share neighbors. For instance, element $(3, 5)$ in the upper half of Table 7.8 is both a neighbor of element $(3, 4)$ and of element $(4, 5)$, hence it is updated twice; overall the 8 new values in the band result from 12 updates. Line 22 in Algorithm 7.4 removes duplicates from the active band list.

While the FIM of Algorithm 7.4 introduces independent tasks that can be computed in parallel at each iteration, it is complicated by having nested tasks (lines 6 and 11) and suffers from duplicate computation. A simple change to the algorithm results in two separate levels of tasks and no duplicate work. Algorithm 7.5 shows that we can postpone the updating of neighbor points (line 11 in Algorithm 7.4) by first adding them to another list, removing duplicates, then updating them.

Note that the algorithms we've given for the FIM could easily be modified to only march the band to a desired thickness, as in the sequential FMM.

Domain Decomposed Fast Marching Method

The degree of concurrency of the Fast Iterative Method is limited by the number of active band elements at each iteration. The FIM does away with the priority queue of the FMM to allow independent updates of band elements. An alternative approach is to essentially preserve the sequential FMM, but perform a data decomposition of the matrix. Each task performs the sequential FMM on its subdomain. Of course, now tasks must communicate

Table 7.10: Solution to domain initialized in Table 7.9 with FIM. Values in active band at each iteration are shown in bold.

1.7071	1	**1.7071**	100	100	100	100
1	0	1	2	100	100	100
1.7071	1	**1.7071**	100	100	100	100
100	2	100	100	2	100	100
100	100	100	**1.7071**	1	**1.7071**	100
100	100	2	1	0	1	2
100	100	100	**1.7071**	1	**1.7071**	100

1.7071	1	1.7071	**2.5453**	100	100	100
1	0	1	2	**3**	100	100
1.7071	1	1.7071	**2.5453**	**3**	100	100
2.5453	2	**2.5453**	**2.5453**	2	**2.5453**	100
100	**3**	**2.5453**	1.7071	1	1.7071	**2.5453**
100	**3**	2	1	0	1	2
100	100	**2.5453**	1.7071	1	1.7071	**2.5453**

1.7071	1	1.7071	2.5453	**3.4422**	100	100
1	0	1	2	3	**4**	100
1.7071	1	1.7071	2.5453	**2.9251**	**3.4422**	100
2.5453	2	2.5453	2.5453	2	2.5453	**3.2524**
3.4422	**2.9251**	2.5453	1.7071	1	1.7071	2.5453
4	3	2	1	0	1	2
100	**3.4422**	2.5453	1.7071	1	1.7071	2.5453

1.7071	1	1.7071	2.5453	3.4422	**4.3709**	100
1	0	1	2	**2.9974**	**3.8928**	100
1.7071	1	1.7071	2.5453	2.9251	**3.4163**	**4.0480**
2.5453	2	2.5453	2.5453	2	2.5453	3.2524
3.4163	2.9251	2.5453	1.7071	1	1.7071	2.5453
3.8928	**2.9974**	2	1	0	1	2
4.3709	3.4422	2.5453	1.7071	1	1.7071	2.5453

1.7071	1	1.7071	2.5453	**3.4414**	**4.3378**	100
1	0	1	2	2.9974	**3.8822**	100
1.7071	1	1.7071	2.5453	2.9251	3.4163	**4.0367**
2.5453	2	2.5453	2.5453	2	2.5453	3.2524
3.4163	2.9251	2.5453	1.7071	1	1.7071	2.5453
3.8822	2.9974	2	1	0	1	2
4.3378	**3.4414**	2.5453	1.7071	1	1.7071	2.5453

1.7071	1	1.7071	2.5453	3.4414	**4.3337**	100
1	0	1	2	2.9974	3.8822	**4.6624**
1.7071	1	1.7071	2.5453	2.9251	3.4163	4.0367
2.5453	2	2.5453	2.5453	2	2.5453	3.2524
3.4163	2.9251	2.5453	1.7071	1	1.7071	2.5453
3.8822	2.9974	2	1	0	1	2
4.3337	3.4414	2.5453	1.7071	1	1.7071	2.5453

1.7071	1	1.7071	2.5453	3.4414	4.3337	**5.1858**
1	0	1	2	2.9974	3.8822	4.6624
1.7071	1	1.7071	2.5453	2.9251	3.4163	4.0367
2.5453	2	2.5453	2.5453	2	2.5453	3.2524
3.4163	2.9251	2.5453	1.7071	1	1.7071	2.5453
3.8822	2.9974	2.0000	1	0	1	2
4.3337	3.4414	2.5453	1.7071	1	1.7071	2.5453

Algorithm 7.5: Modified Fast Iterative Method.

Input: Grid spacing h, $(n_i + 2) \times (n_j + 2)$ speed function matrix F,
$(n_i + 2) \times (n_j + 2)$ solution matrix U initialized to large value everywhere
except boundary (source of front), where it is initialized to 0.
Output: Solution matrix U
Data: Copy V of solution matrix U, initialized to same values.
Data: Three lists: L_1, L_2, containing elements in active band, and L_3 containing
candidate elements

1: **foreach** *element (i, j) of U which is a neighbor of a boundary element* **do**
2:     add (i, j) to L_1
3: **end**
4: **while** *L_1 is not empty* **do**
5:     **foreach** *element (i, j) in L_1* **do**
6:        $u_{\text{new}} \leftarrow$ solveQuadratic(U, i, j, F, h) // same function as Alg. 7.1
7:        $V[i, j] \leftarrow u_{\text{new}}$
8:        **if** $|u_{\text{new}} - U[i, j]| < \epsilon$ **then**
9:           **for** $(l, m) \leftarrow (i+1, j)$, $(i-1, j)$, $(i, j+1)$, $(i, j-1)$ **do**
10:             **if** (l, m) *not in* L_1 **then**
11:                add (l, m) to L_3
12:             **end**
13:           **end**
14:        **else**
15:           add (i, j) to L_2 // keep in list for next iteration if not
                     converged
16:        **end**
17:     **end**
18:     remove duplicates from L_3
19:     **foreach** *element (i, j) in L_3* **do**
20:        remove (i, j) from L_3
21:        $u_{\text{new}} \leftarrow$ solveQuadratic(U, i, j, F, h)
22:        **if** $u_{\text{new}} < U[i, j]$ **then**
23:           $V[i, j] \leftarrow u_{\text{new}}$
24:           add (i, j) to L_2
25:        **end**
26:     **end**
27:     swap references to L_1 and L_2
28:     swap references to U and V
29: **end**

---

information periodically, and deal with values on subdomain borders that may not be consistent with its neighboring tasks. Let's take the example from Table 7.9 and decompose it into four subdomains shown in Figure 7.11, including layers of ghost elements. In this case two subdomains updated the neighboring elements of their boundary points while the other two subdomains had no work to do. This points to potential problems of load imbalance for this method. As fronts propagate across subdomains then border values need to be communicated.

Table 7.11: Data decomposition of FMM, including ghost elements (in *italics*): after initialization and exchange of inter-subdomain border elements.

100	1	100	*100*	*100*	100	100	100	100
1	0	1	*100*	*1*	100	100	100	100
100	1	100	*100*	*100*	100	100	100	100
100	*100*	*100*	*100*	*100*	*100*	*100*	*100*	*100*
100	*1*	*100*	*100*	*100*	*100*	*100*	*100*	*100*
100	100	100	*100*	*100*	100	100	100	100
100	100	100	*100*	*100*	100	1	100	100
100	100	100	*1*	*100*	1	0	1	100
100	100	100	*100*	*100*	100	1	100	100

The domain decomposed FMM proposed by Yang and Stern [75] retains much of the sequential FMM, but requires careful treatment of subdomain border elements. It marches the band in stages interrupted by sharing of subdomain border values. The usual domain decomposition is applied, with each subdomain including a layer of ghost elements. However, unlike our Game of Life example, here each task updates ghost elements along with the rest of the subdomain. Therefore this can be called an *overlapping* domain decomposition method. In each iteration each task performs the sequential FMM to a given thickness of the band, exchanges ghost elements that have changed with neighboring tasks and updates its band elements based on values received. Ghost elements are only communicated if they have been updated, unlike other domain decomposition applications where all the inter-subdomain border elements are exchanged.

In this algorithm the only elements that are truly KNOWN are those that are on the interface boundary. All others are provisional, since they may change upon receiving values from other subdomains. Therefore the KNOWN label is replaced with three labels: KNOWN-FIX for values that cannot change, KNOWN-NEW for an element that has been removed from the local priority queue and KNOWN-OLD for an element whose value has been communicated with another subdomain. This last category is used to avoid communicating the same information multiple times to other tasks. The BAND label is replaced by BAND-NEW for FAR elements that have been promoted to the band and are in the priority queue, and BAND-OLD for elements in the queue whose value has been communicated with another task.

Let's now proceed through the stages of the algorithm, given in Algorithms 7.6-7.8. It's not as intimidating as it looks, as much is adapted from the FMM with two new procedures to handle communication and integration of communicated values. The first stage where neighbors of boundary elements (labeled now with KNOWN-FIX) are updated is unchanged (lines 1–3 in Algorithm 7.8) from the sequential Algorithm 7.2, except that tasks with no KNOWN-FIX elements in their subdomain will have nothing to do. The `updateNeighbors` procedure itself has one slight change, in that an element neighboring (i, j) is updated only if it is not already KNOWN-FIX and its current value is greater than $U[i, j]$ (Algorithm 7.6). This last condition is unnecessary in the sequential FMM since neighbors of a KNOWN point

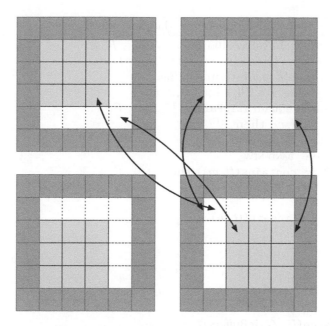

Figure 7.7: FMM 2×2 domain decomposition of a 6×6 grid, showing selected communication. Updates take place on light grey inner elements and on white ghost elements. Communication is two-way: not only are ghost elements received from neighbors, but they are also sent to neighbors. Dark gray elements are for exterior boundary conditions, as in sequential FMM

that aren't KNOWN are all downwind (recall that values downwind are larger). For the parallel FMM this condition allows a KNOWN-OLD or a KNOWN-NEW element to be updated if it is downwind. The solveQuadratic procedure is also slightly changed in Algorithm 7.7 to ensure that only upwind elements are included, since in the parallel FMM KNOWN-NEW and KNOWN-OLD neighbors aren't guaranteed to be upwind.

The while loop used to march the band in marchBand is similar to the one in Algorithm 7.2, with two changes. First, the marching only goes until the thickness reaches a bound, since it needs to be stopped so that ghost elements can be exchanged. Second, the label of the element taken from the queue is changed to KNOWN-OLD or KNOWN-NEW depending on whether it was a BAND-OLD or BAND-NEW element.

So far, the changes to the sequential FMM have been minimal. The parallel FMM introduces two new procedures to gather KNOWN-NEW and BAND-NEW elements that are shared with other tasks (in ghost layer and adjacent layer), send them to neighboring tasks, and integrate received values into each subdomain (see Algorithm 7.6). An example domain decomposition, illustrating two of the exchanges, is shown in Figure 7.7. The integrate procedure updates values of elements received from neighbors, counts the number of updates of KNOWN elements, and updates the queue. An element received from another subdomain is added to the queue if it has a lower value than the local element. The labeling of values outside the bound as BAND-OLD is to maintain consistent labeling across subdomains of elements on either side of the bound.

The main algorithm in Algorithm 7.8 clearly shows how the control flow differs from the sequential FMM. Instead of a single march to the desired width, here the march occurs in stages. The band is marched to a distance given by stride, neighboring values are exchanged

---

Algorithm 7.6: Procedures for Parallel Fast Marching Method.

Procedure updateNeighborsPar(iQmin, U, G, i, j, F, h)

    for $(l, m) \leftarrow (i+1, j), (i-1, j), (i, j+1), (i, j-1)$ **do**

        if (l, m) *not in subdomain* **then** continue

        if $G[l, m] \neq$ KNOWN-FIX $\wedge U[l, m] > U[i, j]$ **then**

            $u_{\text{temp}} \leftarrow$ solveQuadraticPar(G, U, l, m, F, h)

            if $u_{temp} < U[l, m]$ **then**

                $U[l, m] \leftarrow u_{\text{temp}}$

                $G[l, m] \leftarrow$ BAND-NEW

                if iQmin.contains((l, m)) **then**

                    iQmin.change($(l, m), u_{temp}$)

                else

                    iQmin.insert($(l, m), u_{temp}$)

                end

            end

        end

    end

end

Procedure exchange(U, G, inBuffers)

    foreach (i, j) *in or adjacent to ghost region* **do**

        if $G[i, j] =$ BAND-NEW $\vee G[i, j] =$ KNOWN-NEW **then**

            if $G[i, j] =$ BAND-NEW **then**

                $G[i, j] \leftarrow$ BAND-OLD

            else

                $G[i, j] \leftarrow$ KNOWN-OLD

            end

            add (i, j) and $U[i, j]$ to **outBuffers** for each neighboring subdomain

        end

    end

    send/receive **outBuffers** to neighboring subdomain **inBuffers**

end

Procedure integrate(iQmin, U, G, inBuffers, *count*, bound)

    count $\leftarrow 0$

    foreach $u_{l,m}^{new}$ *in* inBuffers **do**

        if $u_{l,m}^{new} < U[l, m]$ **then**

            if $G[l, m] =$ KNOWN-NEW $\vee G[l, m] =$ KNOWN-OLD **then** *count* $\leftarrow count + 1$

            $U[l, m] \leftarrow u_{l,m}^{new}$

            if $U[l, m] >$ bound **then**

                $G[l, m] \leftarrow$ BAND-OLD

            else

                $G[l, m] \leftarrow$ KNOWN-OLD

            end

            if iQmin.contains((l, m)) **then**

                iQmin.change($(l, m), U[l, m]$)

            else

                iQmin.insert($(l, m), U[l, m]$)

            end

        end

    end

end

Algorithm 7.7: Quadratic solver for Parallel Fast Marching Method.

Procedure solveQuadraticPar(G, U, i, j, F, h)

    $a \leftarrow$ selectMin(G, U, i, j, $i+1$, j, $i-1$, j)

    $b \leftarrow$ selectMin(G, U, i, j, i, $j+1$, i, $j-1$)

    if $a = -1$ **then**

        return $b + h/F[i,j]$

    else if $b = -1$ **then**

        return $a + h/F[i,j]$

    else if $|a - b| \geq h/F[i,j]$ **then**

        return $\min(a,b) + h/F[i,j]$

    else

        return $\left(a + b + \sqrt{2 * (h/F[i,j])^2 - (a-b)^2} \right) / 2$

    end

end

// known includes KNOWN-FIX, KNOWN-NEW, KNOWN-OLD

Procedure selectMin(G, U, i, j, l, m, p, q)

    $x \leftarrow -1$

    if $G[l,m] =$ known $\land U[l,m] < U[i,j] \land G[p,q] =$ known $\land U[p,q] < U[i,j]$ **then**

        $x \leftarrow \min(U[l,m], U[p,q])$

    else if $G[l,m] \neq$ known $\land G[p,q] =$ known $\land U[p,q] < U[i,j]$ **then**

        $x \leftarrow U[p,q]$

    else if $G[l,m] =$ known $\land U[l,m] < U[i,j] \land G[p,q] \neq$ known **then**

        $x \leftarrow U[l,m]$

    end

    return x

end

Algorithm 7.8: Parallel Fast Marching Method.

Input: Grid spacing h, desired width L of solution, **stride** for each march, $(n_i + 2) \times (n_j + 2)$ speed function matrix F, $(n_i + 2) \times (n_j + 2)$ solution matrix U initialized to large value everywhere except boundary (source of front), where it is initialized to 0. U is decomposed into n_p subdomains, each with with a layer of ghost elements, followed by a layer of large values (Figure 7.7).

Output: Solution matrix U

Data: $(n_i + 2) \times (n_j + 2)$ matrix G, with possible values KNOWN-FIX, KNOWN-NEW, KNOWN-OLD, BAND-NEW, BAND-OLD, FAR. Initialized to FAR everywhere except boundary, where it is initialized to KNOWN-FIX. G is decomposed into n_p subdomains the same way as U

Data: Indexed min priority queue iQmin, one for each subdomain

Data: *count* of updated KNOWN-OLD or KNOWN-NEW values, initalized to 0

1: **foreach** *element* (i,j) *of* G *such that* $G[i,j] =$ KNOWN-FIX **do**
2:     updateNeighborsPar(iQmin, U, G, i, j, F, h)
3: **end**
4: **while** *true* **do**
5:     $(l, m) \leftarrow$ iQmin.minIndex()
6:     $u_{\min} \leftarrow \min_{\text{subdomains}}(U[l, m])$
7:     $count_{\max} \leftarrow \max_{\text{subdomains}}(count)$
8:     **if** $u_{min} > L \wedge count_{max} = 0$ **then** break
9:     bound $\leftarrow \min(u_{\min} +$ stride$, L)$
10:     marchBand(iQmin, U, G, F, h, bound)
11:     exchange(U, G, inBuffers)
12:     integrate(iQmin, U, G, inBuffers, *count*, bound)
13:     marchBand(iQmin, U, G, F, h)
14: **end**

15: **Procedure** marchBand(iQmin, U, G, F, h, bound)
16:     **while** iQmin *is not empty* **do**
17:         $(l, m) \leftarrow$ iQmin.minIndex()
18:         **if** $U[l, m] >$ bound **then** break
19:         **if** $G[l, m] =$ BAND-NEW **then**
20:             $G[l, m] \leftarrow$ KNOWN-NEW
21:         **else if** $G[l, m] =$ BAND-OLD **then**
22:             $G[l, m] \leftarrow$ KNOWN-OLD
23:         **end**
24:         iQmin.delMin()
25:         updateNeighborsPar(iQmin, U, G, l, m, F, h)
26:     **end**
27: **end**

---

and integrated, and any new values that have values less than the bound are updated (line 13). The `while` loop terminates when the band has reached the desired thickness L and no KNOWN elements have been updated as a result of values received from neighbors. These conditions require communication between tasks to find the overall extremal values. The value of `stride` is a user-defined parameter, that varies from 0 (one FMM step then communicate) and ∞ (complete entire FMM on subdomain then communicate). If it is too small the communication will be so frequent as to hinder performance, and if it is too large more iterations of the `forall` loop will be required.

To help understand this parallel FMM, let's trace the progress of a band that radiates out from the interface. Initially the band progresses exactly as in the sequential FMM, with the band elements labeled BAND-NEW and each element removed from the queue labeled KNOWN-NEW. Recall that the subdomains are overlapping, so that the local marching includes the elements in the ghost layer. Once the width of the band has increased by the user-defined `stride`, then all elements in the band that are shared with other tasks are sent to neighboring tasks. These elements are also relabeled as BAND-OLD or NEW-OLD so they are not sent again until they are updated. Tasks then examine the elements they've received and update their values in the queue if they have decreased. The marching of the band then needs to be restarted because of updates from the received elements, because some may have values less than the desired bound. In subsequent iterations the process is the same, with the separate accounting of old and new band elements ensuring that the same values aren't communicated twice with neighbors. Elements are labeled as new again if they are updated in `updateNeighborsPar` with a lower value.

Analysis

The iterations of the FIM add work to the sequential FMM, since some elements are computed several times, but the FIM does not have to maintain a priority queue. Therefore the work is of $O(n^2)$. The depth depends on the size of the marching band. If the band is of $O(n)$ size then the depth and the parallelism are both $O(n)$. Scalability will be limited in practice by the size of the list of band elements.

The domain decomposed FMM has the same work as the sequential FMM, $O(n^2 \log n^2)$. The extra work of the parallel FMM is in the handling of values communicated between subdomains, whose number is $O(n)$, which does not change the complexity. The depth depends on the load balance. If the load is balanced, then the depth is $O(n^2 \log n^2 / p)$ and the parallelism is $O(p)$. The depth will increase the more the load is out of balance. In practice the advantages of this algorithm come from the queue sizes being much smaller than the worst case n^2 and low communication overhead because of relatively small messages. It has a potential disadvantage in load imbalance.

7.3   PARALLEL ALGORITHMS

The discussion in the previous section emphasized discovering independent tasks by altering two existing sequential algorithms: FSM and FMM. We were able to examine the advantages and disadvantages of each design, but we were not concerned with machine and programming models. In this section we'll discuss how the designs can be developed into algorithms for the three machine models.

Figure 7.8: Example for indexing of diagonals in SE sweep, Algorithm 7.9. Elements labeled with k from 2 to $n_i + n_j$.

7.3.1   Parallel Fast Sweeping Methods

Leaving aside the decomposition into one task per sweep direction, which scales only to 4 tasks, we've seen two quite different decompositions of the FSM. The first, making use of domain decomposition, could be developed very easily into a message passing algorithm similar to a 2D decomposition of the Game of Life (1D decomposition shown in Algorithm 4.16). After each iteration, each task would determine whether its subdomain converged, and a reduction over all tasks would determine whether global convergence was achieved. This algorithm could easily be transformed into a SPMD shared memory algorithm, where each thread would work on its subdomain and no exchange of messages would be required.

The reordered FSM is a natural candidate for a SIMD or shared memory implementation, as it is a good example of data parallelism, as shown in Algorithm 7.9 [18]. The elements along each diagonal cover indexes $i \in [1..n_i]$ and $j \in [1..n_j]$ (elements at indexes 0 and $n_i + 1$ or $n_j + 1$ are fixed for the external boundary). To see how to calculate the indexes along each diagonal, consider first the sweep in the SE direction, illustrated in the $n_i = 7, n_j = 4$ example in Figure 7.8, where we index the diagonals from $k = 2$ to $n_i + n_j$. Along each diagonal i varies from 1 to $k-1$ until $k > n_j + 1 = 5$ where i starts varying from $k - n_j$ to $k-1$ until $k > n_i + 1 = 8$ at which point i varies from $k - n_j$ to n_i to the end. The corresponding column indexes are $j = k - i$. The axes need to be rotated to index the diagonal elements in the other sweeps. These rotations are parametrized by the a_i, s_i, a_j, s_1, s_2 arguments in `sweepPar`.

The shared memory parallel loops may require some load balancing as the time to execute `solveQuadratic` can vary (see Algorithm 7.1). These parallel loops could also be implemented for SIMD execution.

7.3.2   Parallel Fast Marching Methods

The Fast Iterative Method in Algorithm 7.5 is suitable for shared memory implementation, as it involves independent treatment of items in the list L_1. We have to be careful turning the `foreach` loops into parallel loops, however, to avoid race conditions when creating lists (L_1 for the first loop, L_2 and L_3 for the second, and L_2 for the third). Using simple array lists would lead to race conditions, which would require either mutual exclusion or private copies of the lists which would be merged after executing the parallel loop. The former

---

Algorithm 7.9: Data Parallel Two dimensional Fast Sweeping Method. solveQuadratic defined in Algorithm 7.1.

Input: Grid spacing h, $(n_i + 2) \times (n_j + 2)$ speed function matrix F,
$(n_i + 2) \times (n_j + 2)$ solution matrix U initialized to large value everywhere except boundary (source of front), where it is initialized to 0.

Output: Solution matrix U

while *not converged* **do**
    sweepPar$(U, n_i + 1, -1, -n_i - 1, 1, 1, F, h)$// Northeast
    sweepPar$(U, n_i + 1, -1, n_i + n_j + 2, -1, -1, F, h)$// Northwest
    sweepPar$(U, 0, 1, n_j + 1, -1, 1, F, h)$// Southwest
    sweepPar$(U, 0, 1, 0, 1, -1, F, h)$// Southeast
end

Procedure sweepPar$(U, a_i, s_i, a_j, s_1, s_2, F, h)$
    for $k \leftarrow 2$ *to* $n_i + n_j$ **do**
        $i_1 \leftarrow \max(1, k - nj)$
        $i_2 \leftarrow \min(n_i, k - 1)$
        // i, j, u_{new} are private
        parallel for $i \leftarrow a_i + s_i * i_1$ **to** $a_i + s_i * i_2$ **step** s_i **do**
            $j \leftarrow a_j + s_1 * k + s_2 * i$
            $u_{\text{new}} \leftarrow$ solveQuadratic(U, i, j, F, h)
            if $u_{\text{new}} < U[i, j]$ **then** $U[i, j] \leftarrow u_{\text{new}}$
        end
    end
end

---

solution may result in too much overhead from contention for list updates, and the latter solution would cost overhead for merging lists and removing duplicates. A better approach would be to use a $n_i \times n_j$ matrix for each list, storing 0 or 1 to indicate whether an element is present in the list. This approach eliminates the problem of duplicate elements. Once the parallel loop completed, an array list would be created by traversing the matrix.

The overhead of creating an array list from a sparse matrix can be reduced by doing it in parallel. Each task is assigned a contiguous block of rows, and performs a sum of the non-zero elements. A parallel exclusive prefix sum is then performed of these values, and the resulting array indicates where each task begins writing to the list.

We could also base our parallel FIM on the original Algorithm 7.4. Load balance would be a challenge because of the nested calls to `solveQuadratic` (line 11), which would lengthen considerably the execution times of the corresponding iterations of the `foreach` loop.

Block FIM

The authors of the FIM proposed an alternative algorithm, targeting GPU execution [43]. It coarsens the algorithm by treating blocks of the matrix as elements in the original algorithm, and maintaining a list of active blocks. This approach is suited to the organization of threads into blocks in GPU programs. Algorithm 7.10 uses nested parallelism: the outer parallel loop iterations are assigned to thread blocks, and the inner loop updates the elements of a block using SIMD execution. Because the updates of each element are done independently, they are done n_b times, where n_b is the size of the block, to propagate information across the block (lines 12–20). A Boolean matrix c_b keeps track of the elements that have converged. A SIMD reduction of this matrix determines the overall convergence of the block.

The next step is to update the neighboring blocks of converged blocks to see if any of their values change (lines 21–30). Finally the list of active blocks is re-created based on those that have not converged.

An interesting variation of this algorithm was employed in the March of the Froblins gaming demo [62]. Here the solution of the Eikonal equation was used to simulate the movement of crowds. The grids were small, 128^2 or 256^2, and strong scaling was crucial to achieve real-time performance. They eliminated the convergence checks by performing tests on loops such as lines 12–20 using their speed functions to determine how many iterations were required for convergence. This reduced the number of iterations to less than n_b and eliminated the need for a reduction to determine convergence.

Algorithm 7.10 could be further improved by using the parallel FSM for the inner parallel loops (Algorithm 7.9). This would reduce the number of iterations from n_b to 4.

Parallel FMM

Algorithm 7.8 was developed for distributed memory computers, but it could be implemented for shared memory execution, using the SPMD model. Each thread would work on its private data, and memory copies would be used in place of communication.

Conclusion

Two familiar patterns that we saw in Chapter 6 reoccurred in this chapter. Parallel set operations were used in the Fast Iterative Method and the parallel Fast Marching Method used an iterative refinement SPMD pattern. The Eikonal solvers also provided good examples of techniques used to find parallel decompositions. Sweeping along diagonals, instead of along each row, in the Fast Sweeping Method resulted in independent tasks. The ap-

Algorithm 7.10: Block Parallel Fast Iterative Method for GPUs. Outer parallel loop assigned to thread blocks, and inner parallel loop uses SIMD execution.

Input: Grid spacing h, $(n_i + 2) \times (n_j + 2)$ speed function matrix F, $(n_i + 2) \times (n_j + 2)$ solution matrix U initialized to large value everywhere except boundary (source of front), where it is initialized to 0. Dimensions of 2D decomposition of U that produces $n_{bi} \times n_{bj}$ blocks, each with n_b elements.

Output: Solution matrix U

Data: Copy V of solution matrix U, initialized to same values.

Data: List L of blocks with active elements, stored as nonzero elements of array of length $n_{bi} \times n_{bj}$.

Data: Boolean arrays for convergence status of elements of each block (c_b) and overall for each block(c).

```
 1: parallel for each block b do //label active blocks
 2:     L[b] ← 0
 3:     foreach element (i, j) of U_b do
 4:         if U_b[i + 1, j] = 0 ∨ U_b[i − 1, j] = 0 ∨ U_b[i, j + 1] = 0 ∨ U_b[i, j − 1] = 0 then
 5:             L[b] ← 1
 6:             break
 7:         end
 8:     end
 9: end
10: while L has nonzero elements do
11:     parallel for each block b such that L[b] = 1 do //in list
12:         for i ← 1 to n_b do
13:             parallel for each element (i, j) in block do
14:                 V_b[i, j] ← solveQuadratic(U_b, i, j, F, h)
15:                 if |V_b[i, j] − U_b[i, j]| < ε then c_b[i, j] ← true else c_b[i, j] ← false
16:             end
17:             swap references to U_b and V_b
18:         end
19:         c[b] ← reduction(c_b)
20:     end
21:     parallel for each block b such that L[b] = 0 do //not in list
22:         if b is a neighbor of a converged block then
23:             parallel for each element (i, j) in block do
24:                 V_b[i, j] ← solveQuadratic(U_b, i, j, F, h)
25:                 if |V_b[i, j] − U_b[i, j]| < ε then c_b[i, j] ← true else c_b[i, j] ← false
26:             end
27:             swap references to U_b and V_b
28:         end
29:         c[b] ← reduction(c_b)
30:     end
31:     parallel for each block b do
32:         if c[b] = false then
33:             L[b] ← 1
34:         end
35:     end
36: end
```

	Block FSM Section 7.2.1	Data Par. FSM Alg. 7.9	FIM Alg. 7.5	Block FIM Alg. 7.10	Par. FMM Alg. 7.8
Shared Memory	✓	✓	✓	✓	✓
SIMD /GPU		✓		✓	
Distributed Memory	✓				✓

Table 7.12: Overview of parallel Eikonal solvers and their suitability for machine models.

proach of breaking, or relaxing, dependencies that occur in sequential algorithms provided an increase in the parallelism at the cost of extra work, which was used to produce the Fast Iterative Method.

The parallel Eikonal solvers we surveyed are summarized in Table 7.12, including their suitability for three machine models. Their relative performance depends strongly on the characteristics of the domain, including the geometry of the boundary and the complexity of the speed function. A few tests cases have been used in the literature, including boundary points distributed randomly across the domain and speed functions with complex maze-like barriers. Complex irregular boundaries and speed functions can lead to load imbalance and to the need for an increased number of iterations required for convergence.

Nested algorithms, such as the block FIM, are well suited to GPUs, and they may also be good for shared memory platforms, as they have enhanced locality. Other nested algorithms could be developed, such as combining the block FSM with the diagonal data parallel FSM for each subdomain. The parallel FMM seems to be the best current candidate for distributed memory execution, although load balancing may prove to be a challenge.

7.4 FURTHER READING

James Sethian's book, *Level Set Methods and Fast Marching Methods* [61], provides mathematical background, numerical methods, and applications to computational geometry, fluid mechanics, computer vision, and materials science. Extension of the algorithms to three dimensions is straightforward, and is discussed in the papers referred to in this chapter. Hysing and Turek compare the numerical efficiency and algorithmic complexity of the sequential FSM and FMM, and find that the FMM outperforms the FSM in terms of speed and accuracy [41]. The Eikonal equation can be solved on irregular grids as well as on regular cartesian domains. For example, the authors of the FIM have extended their method to triangulated surfaces [26].

7.5 EXERCISES

There are many practical applications of the Eikonal equation. A suitably chosen application would make a good course project, which would ideally involve either the 3D equation or many calls of a 2D solver as an intermediate step in a larger application in order to justify parallel execution. The goal of the exercises in this section is to enable a deeper understanding of the algorithms and their limitations through implementation and testing.

7.1 Experimentally compare the sequential FMM and FSM algorithms, using carefully chosen cases. How do the characteristics of the boundary and speed function affect the relative merits of the two algorithms?

7.2 Implement the domain decomposed block FSM algorithm using a SPMD shared memory programming model. Study how the number of iterations varies with the number of subdomains.

7.3 Implement the domain decomposed block FSM algorithm using MPI. Report on the scalability for different problem sizes.

7.4 Implement the data parallel FSM Algorithm 7.9 using a shared memory programming model. If available, compare the performance with the block FSM algorithm (Exercise 7.2).

7.5 Implement the FIM Algorithm 7.5 using a shared memory programming model. Study the effect of average band size on performance using appropriate test cases.

7.6 Implement the block FIM Algorithm 7.10 on a GPU using CUDA or OpenCL. Then modify it to do data parallel FSM sweeps within each block. How much of an improvement is obtained?

7.7 Implement the data parallel FSM Algorithm 7.9 on a GPU using CUDA or OpenCL. If available, how does its performance compare to the block FIM (Exercise 7.6, with data parallel FSM sweeps per block)?

7.8 Implement the parallel FMM Algorithm 7.8 using a shared memory programming model. Study the problem of load imbalance for different test cases.

7.9 Implement the parallel FMM Algorithm 7.8 using MPI. Study its scalability for a reasonably large number of cores.

CHAPTER 8

Planar Convex Hull

Computation of the convex hull is a fundamental problem in computational geometry. It allows a set of points in a d-dimensional space to be represented by a polytope (polygon when $d = 2$). It has many applications, such as determining the space occupied by a set of points and detecting the collision of bodies represented by points. Another interesting application is deciding whether a novel event has occurred. Let's say the points in Figure 8.1 represent experimental observations that represent normal behavior, and we make two new observations (gray points). Determining whether these two new points represent novel events depends on whether they are considered to be inside or outside the domain of the experimental points [23]. If we define the domain of the points as their convex hull, then we can see in Figure 8.2 that the lower point can be considered a novel event but the upper point cannot.

We'll focus in this chapter on the planar convex hull problem. A popular analogy is a string stretched around a cluster of nails sticking out of a board. When the string is pulled taught it will lie on the the convex hull of the set of nails.

Definition 8.1 (Planar Convex Hull). *The convex hull of a set of points S on the plane is the smallest convex subset $H \subseteq S$ that contains all the points. H is convex if a line through any pair of points in H lies within the polygon formed by H.*

Parallel algorithms for the convex hull have been studied for many years, perhaps longer than any other problem in computational geometry [19]. They make use of several of the algorithmic and program structures we encountered in Chapters 3 and 4. We'll begin by looking at three sequential algorithms.

8.1 SEQUENTIAL ALGORITHMS

Algorithms for computing the planar convex hull can be classified into those that require points to be sorted by their x coordinate and those that do not. The most popular algorithm that falls into the latter category is *QuickHull*, which as the name suggests bears some resemblance to QuickSort.

QuickHull

QuickHull, listed in Algorithm 8.1, begins by finding the two points p and q with the minimum and maximum x coordinates. Clearly these must be part of the convex hull. The convex hull can be divided into the *upper hull*, containing points above the line between p and q, $\overline{pq}$, and the *lower hull*, containing points below $\overline{pq}$. QuickHull finds the two halves of

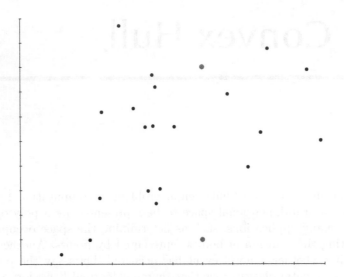

Figure 8.1: Are the gray points novel events?

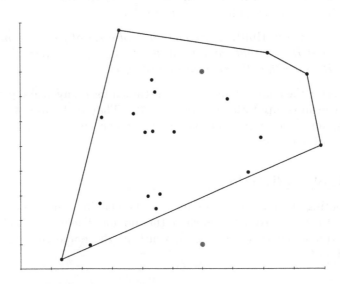

Figure 8.2: Classifying the events represented by gray points using the convex hull. The lower point can be considered novel since it is outside the convex hull.

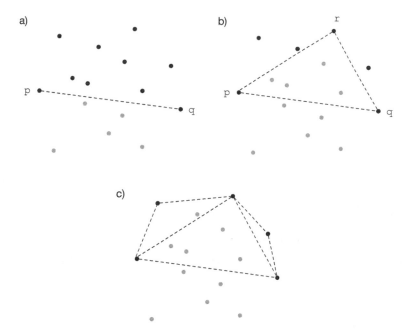

Figure 8.3: Illustration of QuickHull algorithm for upper convex hull. Black points are candidates for the hull and gray points are discarded.

the hull separately. To find the upper hull it discards the points below $\overline{pq}$ (Figure 8.3a), then finds the point r that is the farthest away from $\overline{pq}$. This point is also on the convex hull. The points inside the triangle $\overline{pqr}$ are discarded (Figure 8.3b). The remaining points are partitioned into those that are closest to $\overline{pr}$ or $\overline{rq}$. Then two recursive calls to subHull find the points from these two sets that are on the hull (Figure 8.3c). The recursion terminates when there is no point above $\overline{pq}$. Note that subHull(S, p, q) in Algorithm 8.1 returns the sub-hull delimited by $\overline{pq}$, including the point p but not q.

QuickHull has the same average complexity as QuickSort, $O(n \log n)$, but also the same worst-case complexity, $O(n^2)$. The worst case for QuickHull occurs when the points are distributed on a circle in such a way that all remaining points are always on one side of the splitter r. This is clearly highly unlikely, and in practice QuickHull is one of the best algorithms for computing the planar convex hull.

Several algorithms for finding the planar convex hull require a preprocessing step to sort the points *lexicographically*, which means sorting by x coordinate, and then by y coordinate for points with the same x coordinate. We'll look at two such algorithms, including another divide and conquer algorithm (MergeHull) and an algorithm that incrementally adds points to the hull (Graham scan).

MergeHull

QuickHull and MergeHull are both divide and conquer algorithms. Whereas QuickHull's work is concentrated in the divide phase, in MergeHull the work is performed in the combine phase. This is reflected in Algorithm 8.2, where all the work is contained in the joinHulls procedure. Each recursive level divides the ordered set of points in two and joins the two

---

Algorithm 8.1: QuickHull.

Input: Set S of n points on the plane.
Output: Set H of points on convex hull.

$a \leftarrow \text{minIndex}(\{c_x : c \in S\}) //$ `index of point with minimum` x `value`
$b \leftarrow \text{maxIndex}(\{c_x : c \in S\})$
$p \leftarrow S[a]$
$q \leftarrow S[b]$
$S_1 \leftarrow \{s \in S \mid s \text{ above } \overline{pq}\}$
$S_2 \leftarrow \{s \in S \mid s \text{ below } \overline{pq}\}$
$H_1 \leftarrow \text{subHull}(S_1, p, q) //$ `upper hull`
$H_2 \leftarrow \text{subHull}(S_2, q, p) //$ `lower hull`
$H = H1 \cup H2$

Procedure subHull(S, p, q)
  if $|S| = 0$ **then**
    return $\{p\}$
  end
  $d \leftarrow \text{maxIndex}(\{distance\ between\ c\ and\ \overline{pq} : c \in S\})$
  $r \leftarrow S[d]$
  $S' \leftarrow \{s \in S \mid \text{points not in triangle } \overline{pqr}\}$
  $S_1 \leftarrow \{s \in S' \mid s \text{ closer to } \overline{pr}\}$
  $S_2 \leftarrow \{s \in S' \mid s \text{ closer to } \overline{rq}\}$
  $H_1 \leftarrow \text{subHull}(S_1, p, r)$
  $H_2 \leftarrow \text{subHull}(S_2, r, q)$
  return $H_1 \cup H_2$
end

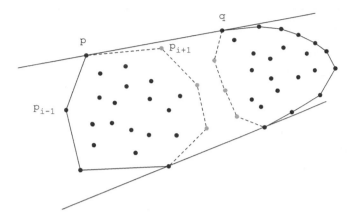

Figure 8.4: Merging of two convex hulls using upper and lower tangents. Tangent defined by points p, q whose left and right neighbors are below the tangent. Gray points in each hull removed from merged hull.

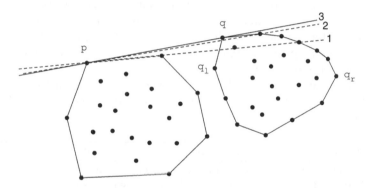

Figure 8.5: Binary search for point q on right hull that forms upper tangent with p. Search is over points on hull from the leftmost point q_l to the rightmost point q_r.

convex hulls. The base case is when there are 3 or fewer points and the convex hull contains all points.

Joining the hulls requires finding the upper and lower tangents and removing from the hull the inner points between the tangents, as illustrated in Figure 8.4. A tangent is defined by points p, q whose left and right neighbors (e.g. p_{i-1} and p_{i+1}) are below $\overline{pq}$. Finding the upper tangent points requires searching among both upper hulls. Figure 8.5 illustrates the case where the right tangent point q, for a given point p on the left hull, is found by a binary search of the points in the upper hull (between q_l and q_r). The tangent point q is found if its left and right neighbors are below the line $\overline{pq}$. Otherwise, the search continues in the direction of the neighbor that is above $\overline{pq}$. We could find both tangent points by doing a binary search in the left upper hull, where at each step a binary search is performed in the right upper hull. This results in an $O(\log|H_1| \log|H_2|)$ algorithm. However, it is possible to do better, using a two dimensional binary search in time $O(\log|H_1| + \log|H_2|)$, which we won't describe here [53]. Removing the points that don't belong on the merged hull takes $O(n)$ time, so the overall complexity is $O(n \log n)$, the same as for QuickHull.

Algorithm 8.2: MergeHull.

Input: Set S of n points on the plane, sorted in lexicographic order.
Output: Set H of points on convex hull, in clockwise order.

Procedure mergeHull(S)
  if $|S| \leq 3$ **then**
    return S
  else
    $H_1 \leftarrow$ mergeHull($S[0..|S|/2)$)
    $H_2 \leftarrow$ mergeHull($S[|S|/2..|S|)$)
    return joinHulls(H_1, H_2)
  end
end

Graham Scan

The Graham scan adds points one at a time to the convex hull, and so is called an incremental algorithm. It is based on the observation that three consecutive points (going clockwise) of a convex hull always make a right turn, that is, the middle point is above the line formed by the outer points. Algorithm 8.3 computes the upper hull and Figure 8.6 shows an example of a few steps of the algorithm. The lower hull is computed in the same way, starting from the rightmost point.

The complexity of the Graham scan, $O(n \log n)$, comes from the requirement to sort the points. The scan itself takes only $O(n)$ time, since points are added one at a time and a point can only be discarded once.

Algorithm 8.3: Graham scan to compute the upper convex hull.

Input: Set S of n points on the plane, sorted in lexicographic order.
Output: Set H of points on upper convex hull, in clockwise order.

add $S[0]$ and $S[1]$ to H
for $i \leftarrow 2$ *to* $n - 1$ **do**
  add $S[i]$ to H
  while $|H| > 2$ *and last 3 points in* H *don't make a right hand turn* **do**
    remove second to last point of H
  end
end

8.2 PARALLEL DESIGN EXPLORATION

The most obvious task decomposition for QuickHull and MergeHull is divide-and-conquer (Section 3.5). As with merge sort we need to go beyond simply assigning a task to each recursive instance. The identification of the splitter r in QuickHull (maxIndex in Algorithm 8.1) can be done using a parallel maximum reduction. Identification of QuickHull's sets S', S_1 and S_2 can be done independently for each point, although a prefix sum is needed to determine where to store subset elements (see Section 8.3.1). If we leave the combine phase of MergeHull as it is in the sequential algorithm, it can be done in $O(\log n)$ time. Once the tangents are identified the merging of two hulls can be done independently for each point,

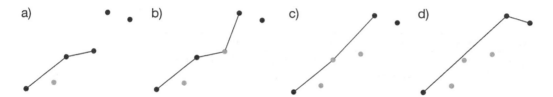

Figure 8.6: Graham Scan example. a) Second point (gray) has already been discarded. b) Adding next point makes a left-hand turn so middle point discarded. c) Last three points of current hull still make a left-hand turn, so middle point discarded. d) adding next point makes a right-hand turn, so existing points kept on provisional hull.

since this only requires writing points that are outside the interval between the tangent points.

These two task decompositions have $O(n \log n)$ work and $O(\log^2 n)$ depth. The depth comes from the $O(\log n)$ levels of $O(\log n)$ scan for QuickHull and $O(\log n)$ computation of the tangents for MergeHull. In practice, the choice between these two algorithms depends on whether the points are sorted. If they are not sorted, we could sort them in parallel. Sorting wouldn't affect the work, but it would change the depth (to $O(\log^3 n)$ for merge sort). The choice of algorithm also depends on the programming model to be used, as we'll see in Section 8.3.

8.2.1   Parallel Hull Merge

We can reduce the depth of MergeHull even further by re-examining the merging of two upper hulls. Assuming each upper hull has m points for convenience, we could decompose the merge into m^2 tasks by examining all possible pairs of points, each pair consisting of a point from each upper hull. This would not be work optimal since as we saw above the merge can be done in $O(\log m)$ time. A better approach is to use this brute force solution, but instead do it over every $\sqrt{m}$th point from each hull, and use m tasks. Of course this isn't the entire solution, as the tangent points aren't necessarily among our selected points. Doing this correctly takes some care [3].

Let's start by looking at the problem of finding the tangent between a point and an upper hull, where the point is to one side of the hull. We've already seen in Figure 8.5 how this can be done sequentially with a binary search. We can decompose this problem in two steps, as shown in Algorithm 8.4. First we use $\sqrt{m}$ tasks, each one determining whether point $q_{i\sqrt{m}}$ is on the tangent by checking whether $q_{(i-1)\sqrt{m}}$ and $q_{(i+1)\sqrt{m}}$ are both below $\overline{pq_{i\sqrt{m}}}$ (line 6). There are also two boundary cases when the tangent is in among the first $\sqrt{m}$ points (line 4) or if it is the last point (line 5). One task will have the answer, q_j. If we're lucky q_j will be the tangent point on the whole upper hull. Otherwise the actual tangent point is in the interval $(j - \sqrt{m} \mathinner{\ldotp\ldotp} j)$ if $q_{j-\sqrt{m}}$ is above $\overline{pq_j}$, and in the interval $(j \mathinner{\ldotp\ldotp} j + \sqrt{m})$ otherwise. Then $\sqrt{m} - 1$ tasks can each determine whether one point in the appropriate interval is on the tangent (lines 16–19). Note that only one task in each of the **parallel for** loops in Algorithm 8.4 will assign a value to j. Figure 8.7 gives an example for $m = 16$.

It would be convenient for our merge to be able to select $\sqrt{m}$ points on each hull, find the tangent, and search in the intervals to the left and right of each tangent point. However, it's not hard to construct an example where one of the actual tangent points is far from the tangent point of the subset of the same hull. One of the tangent points is guaranteed to be in one or both of the neighboring intervals of the subset tangent, however [3]. We can

---

Algorithm 8.4: Tangent between point and upper hull.

Input: Point p to left or right of upper convex hull H^u on the plane, $|H^u| = m$.
Assume m is square for convenience. Points $[q_1..q_m]$ in H^u stored in clockwise order.

Output: Index of point in H^u that forms tangent with p.

1: **Procedure** findPointHullTangent(p, H^u)
2:    $t \leftarrow \sqrt{m}$
3:    **parallel for** $i \leftarrow 1$ *to* t **do** //find index of tangent among every t points
4:       **if** $i = 1 \wedge q_{2t}$ *below* $\overline{pq_t}$ **then** $j \leftarrow t$
5:       **else if** $i = t \wedge q_{m-t}$ *below* $\overline{pq_m}$ **then** $j \leftarrow m$
6:       **else if** $q_{(i-1)t}$ *and* $q_{(i+1)t}$ *below* $\overline{pq_{it}}$ **then** $j \leftarrow i * t$
7:    **end**
8:    **if** $j = m$ **then return** m
9:    **if** q_{j-1} *below* $\overline{pq_j}$ **then**
10:       $l \leftarrow j + 1$
11:       $r \leftarrow j + t - 1$
12:    **else**
13:       $l \leftarrow j - t + 1$
14:       $r \leftarrow j - 1$
15:    **end**
16:    **parallel for** $i \leftarrow l$ *to* r **do**
17:       **if** $i = 1 \wedge q_2$ *below* $\overline{pq_1}$ **then** $j \leftarrow 1$
18:       **else if** q_{i-1} *and* q_{i+1} *below* $\overline{pq_i}$ **then** $j \leftarrow i$
19:    **end**
20:    **return** j
21: **end**

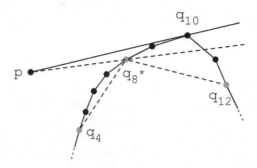

Figure 8.7: Finding tangent between p and upper hull. Tangent point q_8^* found among subset of every 4 points (gray). q_7 is below this tangent, so the actual tangent point is among $[q_9, q_{10}, q_{11}]$. This turns out to be q_{10}.

bracket each interval by finding the tangent from each of the endpoints, as illustrated in Figure 8.8.

Procedure `findInterval` in Algorithm 8.5 determines whether $p_i \in H_1^u$ is on the tangent of H_1^u and H_2^u (then we're done), whether the left tangent point is in the interval to the left or right of p_i, or whether it is somewhere else in H_1^u. `findInterval` makes us of procedure `tangentCompare` (Algorithm 8.6) to determine whether the tangent point is to the left or the right of a point in H_1^u. `tangentCompare` does this by calling `findPointHullTangent` (Algorithm 8.4) to find the tangent between p_i and H_2^u and examining the points p_{i-1} and p_{i+1}.

Algorithm 8.5: Procedure `FindInterval`

Input: Convex upper hulls H_1^u and H_2^u, $p_i \in H_1^u$, index i of p_i. H_1^u to the right or left of H_2^u. Assume $|H_1^u|$ is square for convenience.

Output: Returns status flag f followed by 2 indices. Returns $f = 0$ if both tangent points found, followed by their indices. Returns $f = -1$ if tangent point in neither interval, and indices (0,0). Otherwise returns $f = 1$ and beginning and end indices of interval of size $\sqrt{|H_1^u|}$ on side of p_i containing tangent point.

Procedure `findInterval`(p_i, H_1^u, H_2^u)

$\quad s \leftarrow \sqrt{|H_1^u|}$

$\quad (q_\tau, a) \leftarrow$ `tangentCompare`(p_i, H_1^u, H_2^u)

$\quad$ **if** $a = 0$ **then return** $(0, i, \tau)$

$\quad$ **if** $a = -1$ **then**

$\quad\quad$ **if** $i = s$ **then** //tangent among first $s-1$ points of $H_1^{u'}$

$\quad\quad\quad$ **return** $(1, 1, s-1)$

$\quad\quad$ **else**

$\quad\quad\quad (q_\sigma, b) \leftarrow$ `tangentCompare`(p_{i-s}, H_1^u, H_2^u)

$\quad\quad\quad$ **if** $b = 0$ **then return** $(0, i-s, \sigma)$

$\quad\quad\quad$ **if** $b = 1$ **then**

$\quad\quad\quad\quad$ **return** $(1, i-s+1, i-1)$

$\quad\quad\quad$ **end**

$\quad\quad$ **end**

$\quad$ **else**

$\quad\quad (q_\sigma, b) \leftarrow$ `tangentCompare`(p_{i+s}, H_1^u, H_2^u)

$\quad\quad$ **if** $b = 0$ **then return** $(0, i+s, \sigma)$

$\quad\quad$ **if** $b = -1$ **then**

$\quad\quad\quad$ **return** $(1, i+1, i+s-1)$

$\quad\quad$ **end**

$\quad$ **end**

$\quad$ **return** (-1,0,0)

end

The hull merge Algorithm 8.7 [3] considers several cases in turn. It starts by finding the tangent $\overline{p_{i'}q_{j'}}$ between the coarse hulls C_1^u and C_2^u (lines 5–7). Next it checks whether the actual tangent points are in the intervals adjacent to $p_{i'}$ and $q_{j'}$ (lines 9,11). There are three occasions where it finds the tangent early (line 8, 10, 12). Then it considers three cases:

1. Lines 13–18: Both tangent points are in adjacent intervals (as in Figure 8.8)

2. Lines 19–28: The left tangent point is outside the intervals next to $p_{i'}$

Algorithm 8.6: Procedure `TangentCompare`

Input: Convex upper hulls H_1^u and H_2^u, $p_i \in H_1^u$.
Output: returns tangent q_j on H_2^u and (0, 1 or -1): 0 if p_i is on tangent, -1 if
        tangent point of H_1^u is to the left of p_i, +1 if it is to the right.
Procedure tangentCompare(p_i, H_1^u, H_2^u)
    $q_j \leftarrow$ findPointHullTangent(p_i, H_2^u)
    if p_{i-1} *and* p_{i+1} *below* $\overline{p_i q_j}$ **then**
        return $(q_j, 0)$
    else if p_{i-1} *above* $\overline{p_i q_j}$ **then**
        return $(q_j, -1)$
    else
        return $(q_j, 1)$
    end
end

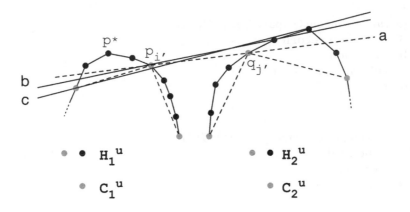

Figure 8.8: Finding tangent $\overline{p^* q^*}$ between upper hulls H_1^u and H_2^u. a) Tangent $\overline{p_{i'} q_{j'}}$ found between coarser hulls C_1^u and C_2^u. b) Tangent between $p_{i'}$ and H_2^u computed using `findPointHullTangent`. $p_{i'-1}$ above this tangent, so the actual tangent point p^* is to the left. c) Tangent between $p_{i'-4}$ and H_2^u computed. $p_{i'-3}$ above this tangent, so p^* is between $p_{i'-4}$ and $p_{i'}$. The same procedure occurs for $q_{j'}$ and H_1^u. This case where both p^* and q^* are adjacent to $p_{i'}$ and $p_{j'}$ respectively corresponds to case at lines 13–18 of Algorithm 8.7.

Algorithm 8.7: Tangent of two upper hulls

Input: Convex upper hulls H_1^u and H_2^u, where H_1^u is to the left of H_2^u. Points $[p_1..p_m]$ in H_1^u and $[q_1..q_n]$ in H_2^u stored in clockwise order. Assume m and n are square for convenience.

Output: Index of point in each of H_1^u and H_2^u that form tangent between both hulls.

1: $s \leftarrow \sqrt{m}$
2: $t \leftarrow \sqrt{n}$
3: $C_1^u \leftarrow [p_s, p_{2s}, p_{3s}, ..p_m]$
4: $C_2^u \leftarrow [q_t, q_{2t}, q_{3t}, ..q_n]$
5: **parallel for** *every pair of points* $p_i \in C_1^u$ *and* $q_j \in C_2^u$ **do**
6:      **if** p_i *and* q_j *on tangent between* C_1^u *and* C_2^u **then** $(i', j') \leftarrow (i, j)$
7: **end**
8: **if** $p_{i'}$ *and* $q_{j'}$ *on tangent between* H_1^u *and* H_2^u **then return** tangent is $\overline{p_{i'}q_{j'}}$
9: $(f_1, l_1, r_1) \leftarrow \texttt{findInterval}(p_{i'}, H_1^u, H_2^u)$
10: **if** $f_1 = 0$ **then return** tangent is $\overline{p_{l_1}q_{r_1}}$
11: $(f_2, l_2, r_2) \leftarrow \texttt{findInterval}(q_{j'}, H_2^u, H_1^u)$
12: **if** $f_2 = 0$ **then return** tangent is $\overline{p_{r_2}q_{l_2}}$
13: **if** $f_1 = 1 \wedge f_2 = 1$ **then**
14:      **parallel for** *every pair of points* $p_i \in H_1^u[l_1..r_1]$ *and* $q_j \in H_2^u[l_2..r_2]$ **do**
15:          **if** p_i *and* q_j *on tangent between* H_1^u *and* H_2^u **then** $(i', j') \leftarrow (i, j)$
16:      **end**
17:      **return** tangent is $\overline{p_{i'}q_{j'}}$
18: **end**
19: **if** $f_1 = -1$ **then**
20:      **parallel for** *every pair of points* $p_i \in C_1^u$ *and* $q_j \in H_2^u[l_2..r_2]$ **do**
21:          **if** p_i *and* q_j *on tangent between* C_1^u *and* H_2^u **then** $(i', j') \leftarrow (i, j)$
22:      **end**
23:      $(f, l, r) \leftarrow \texttt{findInterval}(p_{i'}, H_1^u, H_2^u)$
24:      **if** $f = 0$ **then return** tangent is $\overline{p_l q_r}$
25:      **parallel for** *every pair of points* $p_i \in H_1^u[l..r]$ *and* $q_j \in H_2^u[l_2..r_2]$ **do**
26:          **if** p_i *and* q_j *on tangent between* H_1^u *and* H_2^u **then** $(i', j') \leftarrow (i, j)$
27:      **end**
28:      **return** tangent is $\overline{p_{i'}q_{j'}}$
29: **else**
30:      **parallel for** *every pair of points* $p_i \in H_1^u[l_1..r_1]$ *and* $q_j \in C_2^u$ **do**
31:          **if** p_i *and* q_j *on tangent between* H_1^u *and* C_2^u **then** $(i', j') \leftarrow (i, j)$
32:      **end**
33:      $(f, l, r) \leftarrow \texttt{findInterval}(q_{j'}, H_2^u, H_1^u)$
34:      **if** $f = 0$ **then return** tangent is $\overline{p_r q_l}$
35:      **parallel for** *every pair of points* $p_i \in H_1^u[l_1..r_1]$ *and* $q_j \in H_2^u[l..r]$ **do**
36:          **if** p_i *and* q_j *on tangent between* H_1^u *and* H_2^u **then** $(i', j') \leftarrow (i, j)$
37:      **end**
38:      **return** tangent is $\overline{p_{i'}q_{j'}}$
39: **end**

---

3. Lines 29–38: The right tangent point is outside the intervals next to $q_{j'}$

All this effort allows us to find the upper tangent between two upper hulls in $O(1)$ depth. The iterations of the **parallel for** loops are all independent, so they have constant depth. Algorithm 8.4 (**findPointHullTangent**) is called up to 6 times (via calls to **findInterval**), and has a **parallel for** loop of constant depth. The parallel loops involve either $\sqrt{m}$ (**findPointHullTangent**) or m tasks, each task doing constant work, so overall the work is $O(m)$ (assuming $m = |H_1^u| = |H_2^u|$ for simplicity). The fact that this algorithm is not work efficient (the best sequential algorithm is $O(\log m)$) is not a problem once we incorporate it into MergeHull. Recall that MergeHull has $O(\log n)$ steps, which means that the work is $O(n \log n)$ and the depth is $O(\log n)$. This gives us a work optimal convex hull algorithm with optimal $O(n)$ parallelism.

An incremental algorithm, such as the Graham Scan, doesn't lend itself to parallel decomposition but can be a useful step when we consider algorithms for some of the machine models.

8.3 PARALLEL ALGORITHMS

Developing a convex hull algorithm for a shared memory fork-join framework not as straightforward as it first appears. The easy part is that both algorithms spawn a task at each level of recursion. QuickHull also requires a parallel reduction and parallel creation of subsets S, S_1 and S_2. MergeHull requires a parallel merge and both algorithms also require parallel assembly of the pieces of the hull.

Let's use the fork-join guidelines of Section 4.3 to frame our discussion. The first guideline instructs us to limit the number of tasks by only spawning one new task at each level. For the second guideline we should use a sequential cutoff to avoid an overabundance of tasks. Tasks are only spawned if the size of the set of points in the argument of **subHull** or **MergeHull** is above the cutoff. Below the cutoff, the same algorithm could continue sequentially, or another sequential algorithm could be used. Using a Graham Scan below the cutoff for parallel MergeSort makes sense, as it only takes $O(n)$ time. This guideline also applies to the parallel loops in Algorithm 8.7 for MergeHull and to the maximum reduction and creation of subsets in QuickHull, where there should be fewer tasks than the number of iterations. These can be done with a divide-and-conquer reduction and with either parallel loop constructs or by spawning chunks of iterations using fork-join.

Applying the third guideline, to limit allocation of memory, is simple for the divide phase of MergeHull, where the same array of points can be used throughout the levels of recursion. The parallel merge of Algorithm 8.7 can use strided array accesses for the coarse hulls (C_1^u and C_2^u), rather than creating new arrays. The allocation of memory in the divide phase of QuickHull (Algorithm 8.1) can't be avoided, as the points in subsets S', S_1 and S_2 are arbitrarily scattered in the original array S. The combine phase of both algorithms requires allocation of memory for each piece of the hull. However, we'll see below that a non-recursive data parallel implementation of QuickHull can limit memory allocation by compacting and permuting elements in the same array.

Finally, for the fourth guideline we have to look for any problems with shared data. Both parallel QuickHull and MergeHull require shared read access to the points. There are no race conditions in either parallel QuickHull, where a parallel reduction safely finds the maximum, nor in parallel MergeHull, where tasks only modify their own data or only one task modifies shared data.

Convex hull algorithms for other programming models make good case studies of non-

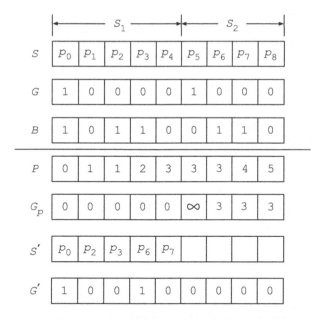

Figure 8.9: Compaction of array S to form subset S' (Algorithm 8.8). Segment arrays G and G' mark the beginning of two segments of array S and S' respectively. Boolean array B indicates elements to be kept to form S'. Compaction produces permutation array P that indicates where elements are to be stored in S' and permutation array G_p to indicate where new segments heads to be placed (elements with $P[i] = G_p[i]$). These arrays are used to produce arrays S' and G'.

recursive divide-and-conquer algorithms. A challenge with a non-recursive approach is that we need a way to identify the segment of the array that each task will work on.

8.3.1 SIMD QuickHull

A non-recursive version of QuickHull is a good candidate for a SIMD algorithm, which is designed for GPU execution. The maximum scan can be done efficiently, and the other operations involve independent updates of array elements. The only complicated part is in the creation of subsets of points. In order to save memory and time we want to be able to do this in place. We can do this using two kinds of data movement [73]: *compaction*, that is, removing points, in the case of creating S', and *split*, that is, reordering the elements of an array according to a label (such as when creating S_1 and S_2).

Compaction

Compaction of an array can easily be decomposed into independent tasks with the help of the familiar scan operation. First, the elements to be kept are independently marked using a separate boolean array (array B in Figure 8.9). An exclusive prefix sum of B produces a permutation array P that indicates where elements are to be stored in the new array. Array P provides the location where independent tasks store their element (if the corresponding element of B is true) in the compacted array S'. Note that we can get the size of the

---

Algorithm 8.8: Segmented array compaction.

Input: Array S of length n, divided into segments according to array G that indicates with a 1 where each segment begins. Array B that indicates with a 0 which elements of S are to be removed.

Output: Permutation arrays P and G_p that can be used to compact S and G, length n of compacted array.

Procedure compact(S, G, B, P, G_p, n)
  $P \leftarrow$ scan(B, sum, exclusive, forward)
  $\{G_p[i] \leftarrow \infty : i \in [0..n)\}$
  $\{T[i] \leftarrow \infty : i \in [0..n)\}$
  $\{T[i] \leftarrow P[i] : i \in [0..n) \mid B[i] = 1\}$
  $G_p \leftarrow$ segmentedScan(T, min, inclusive, forward)
  $n \leftarrow P[n-1] + B[n-1]$
  return n
end

---

compacted array by adding the last element of B to the last element of P, because the final value of the exclusive scan doesn't include the final value in the array.

For divide-and-conquer algorithms we need to be able to operate on disjoint segments of an array. For QuickHull the array initially has two segments, representing the lower and upper hulls. Then these segments are split into smaller segments as the algorithm proceeds. A good way to identify each segment is to use an array of flags, where a value of 1 indicates the beginning of a segment. This is illustrated in array G in Figure 8.9, which demarcates two segments of array S. Compacting array S results in different locations for segments, so we also need to create a new segment array G'.

To find the first index of each new segment we could try finding the value of P corresponding to each old segment boundary, which would give us 0 and 3 for Figure 8.9. However, this wouldn't work if a segment was removed completely. Instead we can do an inclusive min-scan of a temporary array T that is a copy of P with the elements corresponding to zero entries in B replaced with ∞, which is the identity for the min operator. However, we want to scan each segment, not the whole array. This is called a *segmented scan* [8, 58]. For the example in Figure 8.9, a segmented inclusive min scan of $T = [0, \infty, 1, 2, \infty, \infty, 3, 4, \infty]$ results in G_p. The indexes where the new segment boundaries should be stored are given by elements of G_p where $P[i] = G_p[i]$. Algorithm 8.8 [73] performs compaction of a segmented array S, with its segment array G, with information provided by boolean array B.

Split

The divide phase of QuickHull requires each segment of array S to be split in two. Once we determine which new segment each element belongs to we need to permute the array so that each segment occupies a contiguous section. Figures 8.10–8.12 illustrate the process, which is specified in Algorithm 8.9 [8, 73]. Each segment of S (specified by G), is split in two according to the label given in array F. In each segment we want to place the elements of S labeled with a 0 in F followed by those labeled with a 1, and then create new entries in G to mark the new segments.

Let's start with the first half of each split. The location of each element of S that belongs to the first half of each segment is found from a segmented exclusive prefix sum of the complement M of F, which produces array T_M (Figure 8.10). The values in T_M are indexed relative to the beginning of each segment. To compute the offset to the beginning of the array we create a copy of G with the nonzero entries replaced by their index, and take a segmented inclusive max scan of the result, which produces array G_h. This allows us to calculate the entries in the permutation array P corresponding to zero entries in F, given by $T_M + G_h$, as seen in Figure 8.11.

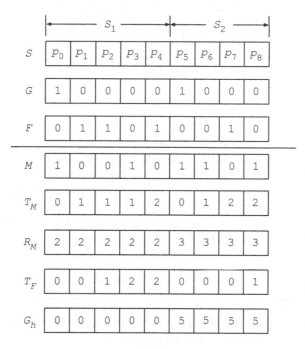

Figure 8.10: Computation of T_M, R_M, T_f, G_h arrays for splitting and permuting segments of S according to F.

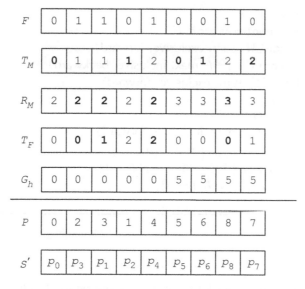

Figure 8.11: Computation of permutation array P, which permutes elements of S to form S'.

	0	1	**2**	3	4	5	6	7	**8**
R_M	2	2	**2**	2	2	3	3	3	**3**
G_h	0	0	0	0	0	5	5	5	**5**
G	1	0	**1**	0	0	1	0	0	**1**

Figure 8.12: Updating segment array G after permutation of S, for those elements where $R_M[i] + G_h[i] = i$.

Creating entries of P for the second half of each split takes more work. We start with a segmented exclusive prefix sum of F, producing T_F. The values of T_F are indices relative to the beginning of each new segment. We need to know the size of the other side of the split so that we can make these indices relative to the beginning of each old segment. This information can be obtained from array T_M, but we need to take a reverse segmented inclusive max scan of $T_M + M$ to produce R_M. A reverse scan is done starting from the end of the array and moving to the beginning. Then we can calculate elements of P corresponding to nonzero elements of F by computing $T_F + R_M + G_h$, as seen in Figure 8.11. We can finally update the segment array G by changing $G[i]$ to 1 if $R_M[i] + G_h[i] = i$ (Figure 8.12).

Algorithm 8.9: Segmented array split and permute.

Input: Array S of length n, divided into segments according to array G that indicates with a 1 where each segment begins. Array F that indicates with a 0 or 1 which side of split each element is on.

Output: Permutation array P and updated G. P is used to permute the elements of S.

Procedure split(S, G, F, P, n)
    $M \leftarrow \neg F$// M is complement of F
    $T_M \leftarrow$ segmentedScan(M, sum, exclusive, forward)
    $R_M \leftarrow$ segmentedScan($T_M + M$, max, inclusive, backward)
    $T_F \leftarrow$ segmentedScan(F, sum, exclusive, forward)
    $\{G_h[i] \leftarrow 0 : i \in [0..n)\}$
    $\{G_h[i] \leftarrow i : i \in [0..n) \mid G[i] = 1\}$
    $G_h \leftarrow$ segmentedScan(G_h, max, inclusive, forward)

    $\{P[i] \leftarrow T_M[i] + G_h[i] : i \in [0..n) \mid F[i] = 0\}$
    $\{P[i] \leftarrow R_M[i] + T_F[i] + G_h[i] : i \in [0..n) \mid F[i] = 1\}$
    $\{G[i] \leftarrow 1 : i \in [0..n) \mid R_M[i] + G_h[i] = i\}$
end

SIMD QuickHull

Now that we know how to perform the data movement required by the QuickHull divide and conquer algorithm, we can formulate a SIMD QuickHull Algorithm 8.10 [73]. As it progresses it removes points from S and adds them to the hull H. It starts with two reductions to find

Algorithm 8.10: SIMD QuickHull. All set operations over $i \in [0..n)$, where n decreases to 0 by end.

Input: Set S of n points on the plane.
Output: Set H of points on convex hull.

1: $\{G[i] \leftarrow 0\}$
2: $G[0] \leftarrow 1$
3: $a \leftarrow$ reduce(S, minIndexX)// Index of point with minimum x value
4: $b \leftarrow$ reduce(S, maxIndexX)
5: $\{B[i] \leftarrow 1\}$
6: $B[a] \leftarrow 0$
7: $B[b] \leftarrow 0$
8: $p \leftarrow S[a]$, $q \leftarrow S[b]$
9: add $S[a]$ and $S[b]$ to H
   // remove p and q
10: $n \leftarrow$ compact(S, G, B, P, G_p, n)// Generates P and G_p
11: $\{S'[P[i]] \leftarrow S[i] \mid B[i] = 1\}$
12: $\{G[i] \leftarrow 0\}$
13: $\{G[G_p[i]] \leftarrow 1 \mid G_p[i] = P[i]\}$
    // Split into lower and upper hull
14: $\{F[i] \leftarrow 0 \mid S'[i] \text{ above } \overline{pq}\}$
15: $\{F[i] \leftarrow 1 \mid S'[i] \text{ below } \overline{pq}\}$
16: split(S', G, F, P, n)// Generates P and updates G
17: $\{S[P[i]] \leftarrow S'[i]\}$
18: $\{P_c[i] \leftarrow p\}$
19: $\{Q_c[i] \leftarrow q\}$
20: **while** $n > 0$ **do**
21: $\quad \{B[i] \leftarrow 1\}$
22: $\quad \{R_c[i] \leftarrow \text{distance between } S[i] \text{ and } \overline{P_c[i]Q_c[i]}\}$
23: $\quad R_c \leftarrow$ segmentedReduce(R_c, max)// result at beginning of each segment
24: $\quad R_c \leftarrow$ segmentedScan(R_c, max, inclusive, forward)
25: $\quad \{B[i] \leftarrow 0 \mid R_c[i] = S[i] \vee S[i] \text{ in triangle } \overline{P_c[i]Q_c[i]R_c[i]}\}$
26: $\quad \{\text{add } S[i] \text{ to } H \mid R_c[i] = S[i]\}$
27: $\quad n \leftarrow$ compact(S, G, B, P, G_p, n)
28: $\quad \{S'[P[i]] \leftarrow S[i]\}$
29: $\quad \{P_c'[P[i]] \leftarrow P_c[i]\}$
30: $\quad \{Q_c'[P[i]] \leftarrow Q_c[i]\}$
31: $\quad \{R_c'[P[i]] \leftarrow R_c[i]\}$
32: $\quad \{G[i] \leftarrow 0\}$
33: $\quad \{G[G_p[i]] \leftarrow 1 \mid G_p[i] = P[i]\}$
34: $\quad \{F[i] \leftarrow 0 \mid S[i] \text{ closer to } \overline{P_c[i]R_c[i]}\}$
35: $\quad \{F[i] \leftarrow 1 \mid S[i] \text{ closer to } \overline{R_c[i]Q_c[i]}\}$
36: $\quad \{P_c'[i] \leftarrow R_c'[i] \mid F[i] = 1\}$
37: $\quad \{Q_c'[i] \leftarrow R_c'[i] \mid F[i] = 0\}$
38: $\quad$ split(S', G, F, P, n)
39: $\quad \{S[P[i]] \leftarrow S'[i]\}$
40: $\quad \{P_c[P[i]] \leftarrow P_c'[i]\}$
41: $\quad \{Q_c[P[i]] \leftarrow Q_c'[i]\}$
42: **end**

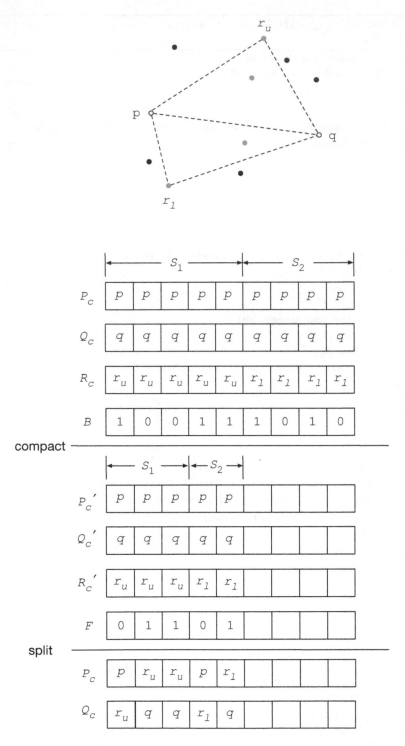

Figure 8.13: One iteration of `while` loop in Algorithm 8.10, showing how arrays that define segment boundaries evolve. Points p and q have already been removed from S. Furthest point from line found for each segment, then it and points inside triangles are removed by compaction (gray points). Remaining points split according to nearest line.

the outer points that define the boundary between the lower and upper hulls. The `compact` procedure is used to remove these points from S (lines 10–13). Next we `split` S into points considered for the upper and lower hulls (lines 14–19). The `while` loop performs the iterative divide and conquer procedure. Arrays P_c and Q_c store the points that define the line for each segment. Figure 8.13 shows how these arrays evolve through the first iteration. There are initially two segments, for the upper and lower hulls. The furthest point from $\overline{pq}$ for each segment (r_u and r_l in Figure 8.13) is computed and stored in R_c (lines 22–24). These points are added to H and removed from S, together with points in each triangle ($\overline{pqr_u}$ and $\overline{pqr_l}$. Note that the same elements are removed from P_c, Q_c, and R_c (lines 25–33). Next each segment is split into those that are closest to $\overline{pr}$ and $\overline{rq}$, and S, P_c and Q_c are permuted according to this split (lines 34–41).

8.3.2 Coarse-Grained Shared Memory MergeHull

We've seen that QuickHull supports a fine grained data parallel decomposition. A combination of independent updates of array elements and scan operations makes Algorithm 8.10 suitable for SIMD execution. On the other hand, MergeHull is suited for coarse grained decomposition. Because the points are sorted it's easy to coarsen the decomposition so that each task has a contiguous chunk of points. Each task can use an efficient sequential algorithm for sorted points, such as the Graham Scan, to compute the convex hull of its points. The interesting part is how to merge the hulls. We'll first develop a shared memory algorithm. Then we'll examine a distributed memory algorithm, where we'll be concerned with the communication overhead.

The high-level Algorithm 8.11 outlines a SPMD shared memory MergeHull. Each thread first performs a Graham scan on a contiguous chunk of points, storing the hulls in array H, with the sizes stored in array N. Then threads cooperate to merge the hulls.

Algorithm 8.11: Shared Memory SPMD MergeHull

Input: Set S of n points on the plane, sorted in lexicographic order.
Output: Set H of points on convex hull.
Data: nt threads, with $id \in [0..nt)$.
Data: Size of nt hulls given by array N.

shared S, H, N
$istart \leftarrow \lfloor id * n/nt \rfloor$
$iend \leftarrow \lfloor (id+1) * n/nt \rfloor - 1$
`grahamScan(`$S, H, N, istart, iend$`)// each hull starts at` $H[istart]$
`hullMerge(`H, N`)`

One way to merge hulls is to iteratively implement the recursive merges, as we have done before with reduction. Algorithm 8.12 merges nt hulls in $\log nt$ steps. At each step $nt/2^i$ pairs of hulls are merged in parallel, with each merge using 2^i threads. The `hullPairMerge` procedure could use an SPMD implementation of Algorithm 8.7 to find the tangents. The assignment of threads to each merge is reminiscent of the parallel merge for merge sort in Algorithm 4.13.

Recall that the main task in merging two hulls is to find the upper and lower tangents. Let's look at an example where we want to merge 4 hulls, in Figure 8.14. Figure 8.15 shows the merging of the upper hulls using Algorithm 8.12. In Figure 8.15a the upper tangents are found between each pair of neighboring hulls, resulting in two upper hulls that are merged

---

Algorithm 8.12: Pairwise hull merge ($nt = 2^i$ threads)

Input: nt hulls stored in H, each hull starting at $H[\lfloor id * n/nt \rfloor]$. Size of each hull in array N.

Output: Merged hull in H

Procedure hullMergeRec(H, N)
    shared H, N
    $istart \leftarrow \lfloor id * n/nt \rfloor$
    $nmt \leftarrow 1$
    for $i \leftarrow 1$ *to* $\log nt$ **do**
        $nmt \leftarrow nmt * 2$
        $idc \leftarrow \lfloor id/nmt \rfloor * nmt$
        // Hulls starting at $H[\lfloor idc * n/nt \rfloor]$ and $H[\lfloor (idc + 2^{i-1}) * n/nt \rfloor]$
        // with sizes $N[idc]$ and $N[idc + 2^{i-1}]$
        hullPairMergePar(H, N, idc, $idc + 2^{i-1}$)
    end
end

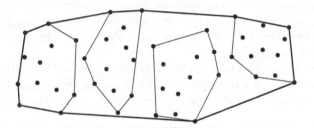

Figure 8.14: Merging 4 hulls.

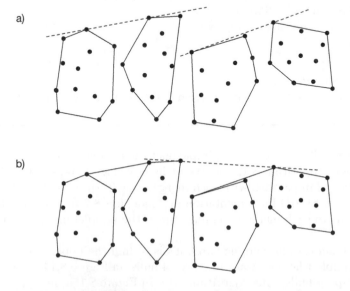

Figure 8.15: Merging 4 upper hulls recursively.

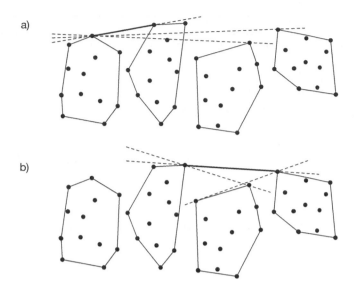

Figure 8.16: Merging 4 upper hulls using tangents between all pairs. a) tangents from the first hull. b) tangents from the second and third hulls. Tangents with greatest slope marked in bold, except for the tangent between the third and fourth upper hulls, which are covered by another tangent.

in Figure 8.15b. Note that none of the points in the third hull are in the final upper merged hull.

An alternative approach is illustrated in Figure 8.16. Here we find the tangent between all pairs of upper hulls. We select the tangent with the largest slope between each hull and all other hulls to its right. If two tangents have the same slope, the one with the right tangent point with the greatest x value is selected. Each of these will define the leftmost and rightmost points p_i^L, p_i^R of each hull that are on the merged hull. Each hull contributes points from p_i^L to p_i^R *unless* the x value of these points are contained in an other $[p_j^R, p_k^L]$ $(j < i, k > i)$ interval, as is the case with the third hull in Figure 8.16. We can decompose this algorithm into a task for the calculation of the tangents from each hull (except the rightmost hull), and tasks to write the points from each hull that are on the merged hull.

Algorithm 8.13 begins with each thread (except thread $id = nt - 1$) finding the tangent between its upper hull and those to the right. The same algorithm applies to the lower hulls, with the starting indexes adjusted appropriately. The $O(\log n)$ sequential algorithm of Overmars and van Leeuwen [53] can be used for findTangent. Then each thread finds the tangent with the maximum slope and populates arrays J, iP and iQ. The id of the right hull of each maximum tangent is stored in J. Arrays iP and iQ store the starting and ending index in H of these tangents. Next each thread identifies the points of it hull that are in the merged hull. First we need a barrier, since we must ensure that all threads have finished writing to arrays J, iP and iQ. For the example in Figure 8.16, $J = [1, 3, 3]$, $iP = [2, 14, 24]$, $iQ = [13, 49, 49]$.

Threads whose hulls are not in the leftmost or rightmost position need to check if their hull is skipped. A hull is skipped if a tangent from a hull to the left has a right endpoint that is to the right of all its points. To identify the starting index *left* we need to find the the leftmost hull whose right tangent endpoint is on the current hull. The ending index *right* is given by the left endpoint of the tangent from the current hull. There are two

---

Algorithm 8.13: SPMD All-pairs upper hull merge.

Input: nt hulls stored in H, each upper hull starting at $H[\lfloor id * n/nt \rfloor]$. Size of each upper hull in array N.

Output: Merged upper hull in H_m

Data: nt threads, with $id \in [0..nt)$.

1: **Procedure** upperHullMergeAllPairs(H, N)
2:     shared H, N, J, iP, iQ
3:     **for** $i \leftarrow id + 1$ *to* $nt - 1$ **do**
4:         $(ip[i], iq[i]) \leftarrow$ findTangent(H, N, id, i)// finds indices in H of tangent
5:     **end**
6:     **if** $id < nt - 1$ **then**
7:         $J[id] \leftarrow$ index i of $\overline{H[ip[i]]H[iq[i]]}$ with maximum slope
8:         $iP[id] \leftarrow ip[J[id]]$
9:         $iQ[id] \leftarrow iq[J[id]]$
10:     **end**
11:     barrier()
12:     $s \leftarrow \lfloor id * n/nt \rfloor$
13:     **if** $id = 0$ **then**
14:         $left \leftarrow 0$
15:         $right \leftarrow iP[0]$
16:     **else if** $id = nt - 1$ **then**
17:         **for** $i \leftarrow 0$ *to* $id - 1$ **do**
18:             **if** $J[i] = id$ **then break**
19:         **end**
20:         $left \leftarrow iQ[i]$
21:         $right \leftarrow s + N[id] - 1$
22:     **else**
23:         $skip \leftarrow 0$
24:         **for** $i \leftarrow 0$ *to* $id - 1$ **do**
25:             **if** $iQ[i] \geq s + N[id]$ **then** $skip \leftarrow 1$
26:         **end**
27:         **if** $skip = 0$ **then**
28:             **for** $i \leftarrow 0$ *to* $id - 1$ **do**
29:                 **if** $J[i] = id$ **then break**
30:             **end**
31:             $left \leftarrow iQ[i]$
32:             $right \leftarrow iP[id]$
33:         **end**
34:     **end**
35:     **if** $skip = 1$ **then** $N[id] \leftarrow 0$ **else** $N[id] \leftarrow right - left + 1$
36:     Offset $\leftarrow$ scan(N, sum, exclusive, forward)
37:     $k \leftarrow 0$
38:     **if** $N[id] \neq 0$ **then**
39:         **for** $i \leftarrow left$ *to* $right$ **do**
40:             $H_m[k + \text{Offset}[id]] \leftarrow H[i]$
41:             $k \leftarrow k + 1$
42:         **end**
43:     **end**
44: **end**

---

border cases, for the first and last hulls, since they have the first and last points of the merged hull. Finally, for threads to write their points to their final positions in parallel, they need to cooperate to do an exclusive prefix sum of N, which contains the number of points contributed from each hull. Then threads write their points to their final positions in H_m.

Analysis

The pairwise merge Algorithm 8.12 seems like a simple and natural solution. The parallel merge of two hulls using Algorithm 8.7 with nt threads takes $O(n/nt)$ time, therefore the overall time is $O((n/nt)\log nt)$. For the all-pairs merge the thread with the first hull needs to find tangents with $nt-1$ other hulls, which takes $O(nt\log(n/nt))$ time. The other steps of Algorithm 8.13 contribute $O(nt+\log nt+n/nt)$, so the overall time is $O(nt\log(n/nt)+n/nt)$. If $n = nt^2$ then the first algorithm has lower complexity. In practice $n >> nt$ for this coarse grained algorithm, in which case the second algorithm will take much less time.

8.3.3   Distributed Memory MergeHull

Let's modify the shared memory SPMD MergeHull of Algorithm 8.11 so that we can create a distributed memory version. We'll assume that the points are already distributed equally among cores. If they are already sorted then each core can perform a Graham Scan to find the hull of its points. Before we proceed further we'll consider what to do if the points are not sorted. Two sorting algorithms that are well suited to distributed data are *bucket sort* and *sample sort*.

Bucket Sort and Sample Sort

If the x values of the points are close to being uniformly distributed then we can use a simple bucket sort. This algorithm assigns a bucket to core id for points in the range $x_{min}+\Delta id/p$ to $x_{min}+\Delta(id+1)/p$, where $\Delta = x_{max}-x_{min}$. Each core sends points that belong to other cores and receives points from other cores that belong in its bucket, then sorts the points in its bucket.

If the points aren't known to be uniformly distributed then *sample sort* modifies bucket sort to ensure that each core gets roughly the same number of values, as listed in Algorithm 8.14 [30]. Sample sort starts with each core sorting its values, and selecting a subset of $p-1$ equally spaced values. These subsets are gathered and sorted by one core, which then selects $p-1$ equally spaced values and broadcasts them to the other cores. Then each core finds the indexes that split the array into p pieces, using a binary search to find each index (other than the beginning and end of the array), and sends subarrays that belong to other cores. Each core merges the chunks it receives from other cores.

The bucket sort takes $O(n/p)$ time to identify chunks and communicate them, and $O((n/p)\log(n/p))$ time to sort. The sample sort takes $O((n/p)\log(n/p))+p^2\log p)$ for both sequential sorts and $O(p\log(n/p))$ to identify chunks using binary search. From Chapter 5 we know that sample sort will take $O(p+\log p)$ for the gather, $O(p\log p)$ for the broadcast, and $O(n/p)$ to communicate chunks. For both sorts, and $n >> p$, the local sequential sorting time dominates. The sample sort achieves a more balanced distribution of values at the cost of only $O(p\log p)$ extra communication time, compared to the bucket sort.

Algorithm 8.14: Sample sort

Input: n values distributed equally among p cores, local values stored in array A.
Output: Sorted values distributed among p cores and according to core id in array C

sequentialSort(A)
$S \leftarrow p - 1$ equally spaced elements of A
gather(0, S, $p - 1$, S)
if $id = 0$ **then**
    sequentialSort(S)
    $B \leftarrow p - 1$ equally spaced elements of S
end
broadcast(0, B, $p - 1$, B)
$iB[0] \leftarrow 0$
for $i \leftarrow 0$ *to* $p - 2$ **do**
    $iB[i + 1] \leftarrow$ binarySearch($A,B[i]$)// first index in A where $B[i] \leq A[i]$
end
$iB[p] \leftarrow |A| - 1$
for $i \leftarrow 0$ *to* $p - 1$ **do**
    if $i = id$ **then continue**
    send $A[iB[i]..iB[i + 1])$ to i
end
$k \leftarrow 0$
$C \leftarrow A[iB[id]..iB[id + 1])$
for $i \leftarrow 0$ *to* $p - 1$ **do**
    if $i = id$ **then continue**
    receive from i into D
    $C \leftarrow$ merge(C, D)
end

Merging Distributed Hulls

The analysis of the hull merge Algorithms 8.12 and 8.13 did not take data movement into account. We can't afford to ignore data movement when the points are distributed over cores on a network. If we applied the all-pairs merge algorithm directly, each core would need to send its complete hull to cores with hulls on the right. To reduce communication we can adopt the idea of working with subsets of the hulls from Algorithm 8.7 and Figure 8.8. Algorithm 8.15, illustrated in Figure 8.17, is adapted from an algorithm of Diallo et al. [19], where we imply use of the $O(\log n)$ hull tangent algorithm [53] instead of a $O(n \log n)$ combination of linear search and binary search. As in all similar algorithms we consider the problem of merging upper hulls.

Figure 8.17a shows that we compute tangents with a subset of the upper hull with all the complete upper hulls to the right. The subset G is created from $p - 1$ equally spaced points of the upper hull H. We first find the tangent $\overline{G[iP]q}$ of G with all complete upper hulls to the right. This is done using procedure findTangentSubset in Algorithm 8.16. Each core sends its subset G to all cores of larger rank. Each core then receives a subset G_{left}, computes the tangent $\overline{G_{left}[iP]q}$ with its upper hull and sends back the index iP of the tangent point in G and the right tangent point q. Each core now has tangents between G and each upper hull to the right. The tangent points of the tangent with the maximum slope are also on the overall merged hull (unless the hull is skipped). findTangentSubset returns these points as the index iP in G and the right tangent point q.

Now we know that the actual tangent point will be in the interval to the left or right of $G[iP]$ (or $G[iP]$ itself if we're lucky). Either the point in H to the left or right of $G[iP]$ will be above $\overline{G_{left}[iP]q}$. Figure 8.17a shows that we create another subset R of the points that are above this line. We use findTangentSubset again to find the point in R that is on the tangent between H and all the upper hulls to the right (point $H[iP]$ in Figure 8.17b).

Finally each core needs to write its points that are on the merged hull to the beginning of its H array. Each core (except $id = p-1$) knows that the last point is $H[iP]$ (Figure 8.17c). The first point is given by the tangent point q with the hull to the left with the smallest x value (except for the leftmost hull). Therefore each core needs to send q (from the tangent $\overline{H[iP]q}$) to all the cores of higher rank. As with the shared memory algorithm, we need to test if the current hull is skipped by the merged hull, which is the case when q from a hull to the left is greater than $H[iP]$, the tangent point to the right.

Analysis

Computation of the local hull takes $O(n/p)$ time. The first call to findTangentSubset involves computing up to $p - 1$ tangent between hulls of size $|G| = p - 1$ and of size $|H| = O(n/p)$, which can be done in $O(p(\log p + \log(n/p))$ time. In the second call the left hulls are of $O(|H|/p) = O(n/p^2)$, but the right hull is of the same size so the time required is $O(p(\log(n/p^2) + \log(n/p))$. Overall the computation time is $O(n/p + p\log(n/p) + p\log p)$. Note that this is the same complexity as the shared memory Merge Hull algorithm if $n > p^2 \log p$.

Most communication takes place in findTangentSubset. Since up to $p - 1$ messages of size $p - 1$ (first call) and $O(n/p^2)$ (second call) are involved, the communication time is $O(p^2 + n/p)$. This is the overall communication time as well, since the only other messages involved are up to $p-1$ messages containing a single point (lines 26 and 29 of Algorithm 8.15).

What if we had simply implemented the all-pairs merge hull without using this two-phase approach? Each core would have had to send all $O(n/p)$ points of its hull to all higher ranked cores, which would require $O(n)$ time, which is considerably worse than Algorithm 8.15.

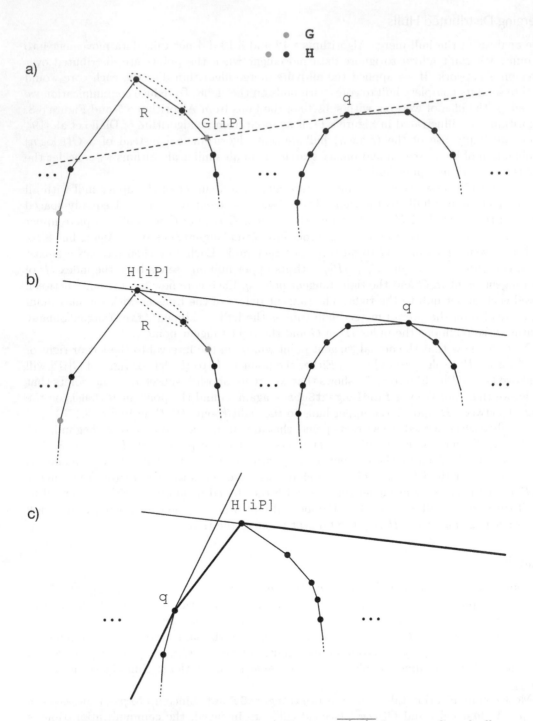

Figure 8.17: Illustration of Algorithm 8.15. a) Tangent $\overline{G[iP]q}$ between G and all upper hulls to the right computed from Algorithm 8.16. Subset R identified, containing points above tangent. b) Tangent $\overline{H[iP]q}$ between R and all hulls to the right computed. c) points on merged upper hull lie between q from tangent on left and $H[iP]$ on tangent on right, inclusive.

Algorithm 8.15: Distributed Memory SPMD Upper Hull Merge

Input: p separate upper hulls, obtained from n lexicographically ordered points partitioned into p chunks. Local hull in $H[0..m]$.

Output: Merged upper hull, with local portion in $H[0..m]$

1: **if** $id < p - 1$ **then**
2:     **for** $i \leftarrow 0$ *to* $p - 2$ **do**
3:         $k \leftarrow \lfloor i * m/p \rfloor, G[i] \leftarrow H[k]$
4:     **end**
5:     $(iP, q) \leftarrow \texttt{findTangentSubset}(G)$
6:     $f \leftarrow 0, k \leftarrow \lfloor iP * m/p \rfloor$
7:     **if** $H[k+1]$ *above* $\overline{G[iP]q}$ **then**
8:         $i1 \leftarrow k + 1, i2 \leftarrow \min(\lfloor (iP+1) * m/p \rfloor, m - 1)$
9:     **else if** $k \neq 0 \wedge H[k-1]$ *above* $\overline{G[iP]q}$ **then**
10:         $i1 \leftarrow \lfloor (iP-1) * m/p \rfloor, i2 \leftarrow k - 1$
11:     **else**
12:         $f \leftarrow 1$ // tangent found
13:     **end**
14:     **if** $f = 0$ **then**
15:         $k \leftarrow 0$
16:         **for** $i \leftarrow i1$ *to* $i2$ **do**
17:             **if** $H[i]$ *above* $\overline{G[iP]q}$ **then**
18:                 $R[k] \leftarrow H[i], k \leftarrow k + 1$
19:             **end**
20:         **end**
21:         $(iP, q) \leftarrow \texttt{findTangentSubset}(R)$
22:     **else**
23:         $iP \leftarrow k$
24:     **end**
25: **end**
26: **for** $j \leftarrow id + 1$ *to* $p - 1$ **do** send q to j
27: $skip \leftarrow 0, first \leftarrow 1$
28: **for** $j \leftarrow 0$ *to* $id - 1$ **do**
29:     receive from j into q_{left}
30:     **if** $q_{left} \geq H[0] \wedge q_{left} \leq H[iP] \wedge first = 1$ **then**
31:         $q \leftarrow q_{left}$
32:         $first \leftarrow 0$
33:     **else if** $q_{left} > H[iP]$ **then**
34:         $skip \leftarrow 1$
35:     **end**
36: **end**
37: $m \leftarrow 0$
38: **if** $id = p - 1$ **then** $iP \leftarrow m - 1$
39: **if** $id = 0$ **then** $q \leftarrow H[0]$
40: **if** $skip = 0$ **then**
41:     **for** $i \leftarrow 0$ *to* iP **do**
42:         **if** $H_x[i] \geq q_x$ **then** $H[m] \leftarrow H[i], m \leftarrow m + 1$
43:     **end**
44: **end**

Algorithm 8.16: Find tangent with hulls to the right.

Procedure findTangentSubset(G)

    for $j \leftarrow id + 1$ *to* $p - 1$ **do**

        send G to j

    end

    for $j \leftarrow 0$ *to* $id - 1$ **do**

        receive from j into G_{left}

        $(iP, iQ) \leftarrow$ findTangent(G_{left}, H)// tangent $\overline{G_{left}[iP]H[iQ]}$

        send iP and $H[iQ]$ to j

    end

    for $j \leftarrow id + 1$ *to* $p - 1$ **do**

        receive from j into $iP[j]$ and $Q[j]$

    end

    $i \leftarrow$ index j of $\overline{H[iP[j]]Q[j]}$ with maximum slope, $j \in (id..p)$

    return $(iP[i], Q[i])$

end

8.4 CONCLUSION

Computation of the planar convex hull is a well understood problem that provides good case studies for divide-and-conquer algorithms for all three machine models. We saw how divide-and-conquer algorithms can be expressed non-recursively, either using segmented SIMD operations or the SPMD pattern for shared and distributed memory implementation. It would be worth exploring the application of the SIMD compaction and splitting procedures to other divide-and-conquer algorithms. The main challenge in the SPMD algorithms was in computing the hull merges, particularly for the distributed memory version where minimizing communication was a concern.

Experimental study of convex hull algorithms needs to take into account different types of data. Worst case analysis may be of interest, where the convex hull includes all points. Typical topologies that should be included are points randomly distributed in a circle or square, or those arising from practical applications, such as creating surfaces from point clouds.

It's important to realize that the convex hull is usually only one part of a larger algorithm. This will guide the selection of programming model and task decomposition. For instance, it may be that points are already sorted, so some variation of MergeHull would be appropriate.

8.5 FURTHER READING

This chapter has only discussed a small subset of existing convex hull algorithms. The standard textbook, *Computational Geometry: Algorithms and Applications* [15], can be consulted. David Mount from the University of Maryland has produced some excellent notes based on this book (http://www.cs.umd.edu/~mount/754/Lects/754lects.pdf). Practical applications, for instance in collision detection or computer vision, are mainly concerned with the 3D convex hull problem. These applications usually employ GPUs, so there have been a number of GPU algorithms proposed [70]. The convex hull problem has been included in the problem based benchmark suite, which is meant to represent non-numerical algorithms [64].

8.6 EXERCISES

8.1 Figure 8.8 is an example of a tangent between hull subsets that has endpoints close to those in the tangent between complete hulls. Give an example where one of the tangent points is far from the tangent point between hull subsets.

8.2 Referring again to the problem illustrated in Figure 8.8, prove that one of the tangent points between hulls must be in the interval next to one of the tangent points between hull subsets.

8.3 Implement a segmented scan in CUDA or OpenCL using a simple algorithm where the scan of each segment is accomplished by a separate thread block. What are the disadvantages of this algorithm?

8.4 Blelloch's scan algorithm (Section 3.4) can be modified to produce a segmented scan. This requires performing logical OR operations on the segment label array. Consult "Scan Primitives for GPU Computing" [58] and express the algorithm using our SIMD set notation. This paper also includes a modified two-level algorithm to make use of the GPU thread hierarchy. Implement this GPU algorithm in CUDA or OpenCL. Compare the performance of this algorithm with the naive version of Exercise 8.3.

8.5 Implement the compact and split procedures (Algorithms 8.8 and 8.9) in OpenCL or CUDA. This will require forward and backward segmented scans (Exercise 8.3 or 8.4).

8.6 A simple radix sort algorithm can be expressed by making use of the split procedure. In each iteration, the integers are split according to the ith bit. Write down this algorithm and implement it for GPU, making use of the split operation from Exercise 8.5.

8.7 Implement and experimentally test the performance of Quickhull Algorithm 8.10, making use of compact and split operations developed in Exercise 8.5.

8.8 Study and implement the hull tangent algorithm of Overmars and van Leeuwen [53].

8.9 Implement and experimentally test the shared memory SPMD MergeHull using all-pairs merge. Algorithm 8.13 can be adjusted to compute the lower hull as well. Use either the optimal $O(\log n)$ tangent algorithm (Exercise 8.8), or a simpler algorithm, such as the $O(\log n \log m)$ algorithm discussed in Section 8.1.

8.10 Implement and experimentally test bucket sort and sample sort (Algorithm 8.14) using MPI. Compare the performance of the two algorithms using suitable test cases (uniform and nonuniform distributions).

8.11 Implement and experimentally test the distributed memory MergeHull of Algorithm 8.15 using MPI, starting from sorted, distributed points. As for Exercise 8.9, use either the optimal $O(\log n)$ tangent algorithm or a simpler algorithm.

Bibliography

[1] Selim G. Akl and Werner Rheinboldt. *Parallel sorting algorithms*. Academic Press, Orlando, 1985.

[2] Gene M. Amdahl. Validity of the Single Processor Approach to Achieving Large Scale Computing Capabilities. In *Proceedings of the April 18–20, 1967, Spring Joint Computer Conference*, AFIPS '67 (Spring), pages 483–485, New York, NY, USA, 1967. ACM.

[3] Mikhail J. Atallah and Michael T. Goodrich. Parallel algorithms for some functions of two convex polygons. *Algorithmica*, 3(1-4):535–548, November 1988.

[4] David H. Bailey, Jonathan M. Borwein, and Victoria Stodden. Facilitating reproducibility in scientific computing: Principles and practice. *Citeseer*, 2014.

[5] Michael Bauer, Sean Treichler, Elliott Slaughter, and Alex Aiken. Legion: Expressing Locality and Independence with Logical Regions. In *Proceedings of the International Conference on High Performance Computing, Networking, Storage and Analysis*, SC '12, pages 66:1–66:11, Los Alamitos, CA, USA, 2012. IEEE Computer Society Press.

[6] Jost Berthold, Mischa Dieterle, Rita Loogen, and Steffen Priebe. Hierarchical Master-Worker Skeletons. In Paul Hudak and David S. Warren, editors, *Practical Aspects of Declarative Languages*, number 4902 in Lecture Notes in Computer Science, pages 248–264. Springer Berlin Heidelberg, January 2008.

[7] G. E. Blelloch. Scans as primitive parallel operations. *IEEE Transactions on Computers*, 38(11):1526–1538, November 1989.

[8] Guy E. Blelloch. *Vector models for data parallel computing*. MIT Press, Cambridge, Mass., 1990.

[9] Guy E. Blelloch and Bruce M. Maggs. Parallel Algorithms. In Mikhail J. Atallah and Marina Blanton, editors, *Algorithms and Theory of Computation Handbook*, pages 25–1 – 25–43. Chapman & Hall/CRC, 2010.

[10] F. Busato and N. Bombieri. An efficient implementation of the Bellman-Ford algorithm for Kepler GPU architectures. *IEEE Transactions on Parallel and Distributed Systems*, PP(99):1–1, 2015.

[11] V. T. Chakaravarthy, F. Checconi, F. Petrini, and Y. Sabharwal. Scalable Single Source Shortest Path Algorithms for Massively Parallel Systems. In *Parallel and Distributed Processing Symposium, 2014 IEEE 28th International*, pages 889–901, May 2014.

[12] Cristian Coarfa, Yuri Dotsenko, John Mellor-Crummey, Franois Cantonnet, Tarek El-Ghazawi, Ashrujit Mohanti, Yiyi Yao, and Daniel Chavarra-Miranda. An Evaluation

of Global Address Space Languages: Co-array Fortran and Unified Parallel C. In *Proceedings of the Tenth ACM SIGPLAN Symposium on Principles and Practice of Parallel Programming*, PPoPP '05, pages 36–47, New York, NY, USA, 2005. ACM.

[13] Thomas H. Cormen, Charles Eric Leiserson, Ronald L. Rivest, and Clifford Stein. *Introduction to algorithms*. MIT Press, Cambridge, Mass., 3rd edition, 2009.

[14] Frederica Darema. SPMD Computational Model. In David Padua, editor, *Encyclopedia of Parallel Computing*, pages 1933–1943. Springer US, 2011.

[15] M. de Berg, M. van Kreveld, M. Overmars, and O. Schwartzkopf. *Computational geometry algorithms and applications*. Springer, Berlin, 3rd edition, 2008.

[16] Mattias De Wael, Stefan Marr, and Tom Van Cutsem. Fork/Join Parallelism in the Wild: Documenting Patterns and Anti-patterns in Java Programs Using the Fork/Join Framework. In *Proceedings of the 2014 International Conference on Principles and Practices of Programming on the Java Platform: Virtual Machines, Languages, and Tools*, PPPJ '14, pages 39–50, New York, NY, USA, 2014. ACM.

[17] Jeffrey Dean and Sanjay Ghemawat. MapReduce: Simplified data processing on large clusters. In *Proceedings of Operating Systems Design and Implementation (OSDI)*, pages 137–150, San Francisco, CA, 2004.

[18] Miles Detrixhe, Frederic Gibou, and Chohong Min. A parallel fast sweeping method for the Eikonal equation. *Journal of Computational Physics*, 237:46–55, March 2013.

[19] Mohamadou Diallo, Afonso Ferreira, Andrew Rau-Chaplin, and Stphane Ubda. Scalable 2d Convex Hull and Triangulation Algorithms for Coarse Grained Multicomputers. *Journal of Parallel and Distributed Computing*, 56(1):47–70, January 1999.

[20] Javier Diaz, Camelia Munoz-Caro, and Alfonso Nino. A Survey of Parallel Programming Models and Tools in the Multi and Many-Core Era. *IEEE Transactions on Parallel and Distributed Systems*, 23(8):1369–1386, August 2012.

[21] J. J. Dongarra. *Sourcebook of parallel computing*. Morgan Kaufmann Publishers, San Francisco, CA, 2003.

[22] Alejandro Duran, Eduard Ayguad, Rosa M. Badia, Jess Labarta, Luis Martinell, Xavier Martorell, and Judit Planas. OmpSs: A Proposal for Programming Heterogeneous Multi-Core Architectures. *Parallel Processing Letters*, 21(02):173–193, June 2011.

[23] T. Ebert, J. Belz, and O. Nelles. Interpolation and extrapolation: Comparison of definitions and survey of algorithms for convex and concave hulls. In *2014 IEEE Symposium on Computational Intelligence and Data Mining (CIDM)*, pages 310–314, December 2014.

[24] Ian Foster. *Designing and building parallel programs: concepts and tools for parallel software engineering*. Addison-Wesley, Reading, Mass., 1995.

[25] Matteo Frigo, Charles E. Leiserson, and Keith H. Randall. The Implementation of the Cilk-5 Multithreaded Language. In *Proceedings of the ACM SIGPLAN 1998 Conference on Programming Language Design and Implementation*, PLDI '98, pages 212–223, New York, NY, USA, 1998. ACM.

[26] Z. Fu, W. Jeong, Y. Pan, R. Kirby, and R. Whitaker. A Fast Iterative Method for Solving the Eikonal Equation on Triangulated Surfaces. *SIAM Journal on Scientific Computing*, 33(5):2468–2488, January 2011.

[27] Vladimir Gajinov, Srdjan Stipi, Igor Eri, Osman S. Unsal, Eduard Ayguad, and Adrian Cristal. DaSH: A benchmark suite for hybrid dataflow and shared memory programming models. *Parallel Computing*, 45:18–48, June 2015.

[28] Michael Garland and David B. Kirk. Understanding Throughput-oriented Architectures. *Commun. ACM*, 53(11):58–66, November 2010.

[29] Sergei Gorlatch. Send-receive Considered Harmful: Myths and Realities of Message Passing. *ACM Trans. Program. Lang. Syst.*, 26(1):47–56, January 2004.

[30] Ananth Grama, Anshul Gupta, George Karypis, and Vipin Kumar. *Introduction to parallel computing*. Addison-Wesley, Harlow, England; New York, second edition, 2003.

[31] William D. Gropp, Ewing Lusk, and Anthony Skjellum. *Using MPI Portable Parallel Programming with the Message-Passing Interface*. MIT Press, Cambridge, Mass., 2014.

[32] Uday G. Gujar and Virendra C. Bhavsar. Fractals from z z + c in the complex c-plane. *Computers & Graphics*, 15(3):441–449, January 1991.

[33] John L. Gustafson. Reevaluating Amdahl's Law. *Communications of the ACM*, 31:532–533, 1988.

[34] Mark Harris. Parallel prefix sum (scan) with CUDA. In *GPU gems 3*, pages 851–876. Addison-Wesley, 2007.

[35] Maurice Herlihy and Nir Shavit. *The art of multiprocessor programming, revised first edition*. Morgan Kaufmann, Waltham, Mass., 2012.

[36] L. Hernndez Encinas, S. Hoya White, A. Martin del Rey, and G. Rodriguez Sanchez. Modelling forest fire spread using hexagonal cellular automata. *Applied Mathematical Modelling*, 31(6):1213–1227, June 2007.

[37] W. Daniel Hillis and Guy L. Steele, Jr. Data Parallel Algorithms. *Commun. ACM*, 29(12):1170–1183, December 1986.

[38] Michelle Hribar, Valerie Taylor, and David Boyce. Performance study of parallel shortest path algorithms: Characteristics of good decompositions. In *Proc. 13th Ann. Conf. Intel Supercomputers Users Group (ISUG)*, 1997.

[39] Michelle Hribar, Valerie Taylor, and David E. Boyce. Parallel Shortest Path Algorithms: Identifying the Factors that Affect Performance. Technical Report CPDC-TR-9803-015, Northwestern University, Evanston, Il., 1998.

[40] Christopher J. Hughes. *Single-Instruction Multiple-Data Execution*. Synthesis Lectures on Computer Architecture. Morgan & Claypool, May 2015.

[41] Shu-Ren Hysing and Stefan Turek. The Eikonal equation: numerical efficiency vs. algorithmic complexity on quadrilateral grids. In *Proceedings of ALGORITMY 2005*, pages 22–31, 2005.

[42] E. Jeannot, G. Mercier, and F. Tessier. Process Placement in Multicore Clusters:Algorithmic Issues and Practical Techniques. *IEEE Transactions on Parallel and Distributed Systems*, 25(4):993–1002, April 2014.

[43] W. Jeong and R. Whitaker. A Fast Iterative Method for Eikonal Equations. *SIAM Journal on Scientific Computing*, 30(5):2512–2534, January 2008.

[44] Kurt Keutzer, Berna L. Massingill, Timothy G. Mattson, and Beverly A. Sanders. A Design Pattern Language for Engineering (Parallel) Software: Merging the PLPP and OPL Projects. In *Proceedings of the 2010 Workshop on Parallel Programming Patterns*, ParaPLoP '10, pages 9:1–9:8, New York, NY, USA, 2010. ACM.

[45] Sara Landset, Taghi M. Khoshgoftaar, Aaron N. Richter, and Tawfiq Hasanin. A survey of open source tools for machine learning with big data in the Hadoop ecosystem. *Journal of Big Data*, 2(1):1–36, November 2015.

[46] Doug Lea. A Java Fork/Join Framework. In *Proceedings of the ACM 2000 Conference on Java Grande*, JAVA '00, pages 36–43, New York, NY, USA, 2000. ACM.

[47] K. Madduri, D. Bader, J. Berry, and J. Crobak. An Experimental Study of a Parallel Shortest Path Algorithm for Solving Large-Scale Graph Instances. In *2007 Proceedings of the Ninth Workshop on Algorithm Engineering and Experiments (ALENEX)*, Proceedings, pages 23–35. Society for Industrial and Applied Mathematics, 2007.

[48] Zoltan Majo and Thomas R. Gross. Memory System Performance in a NUMA Multicore Multiprocessor. In *Proceedings of the 4th Annual International Conference on Systems and Storage*, SYSTOR '11, pages 12:1–12:10, New York, NY, USA, 2011. ACM.

[49] Saeed Maleki, Donald Nguyen, Andrew Lenharth, Mara Garzarn, David Padua, and Keshav Pingali. DSMR: A Shared and Distributed Memory Algorithm for Single-source Shortest Path Problem. In *Proceedings of the 21st ACM SIGPLAN Symposium on Principles and Practice of Parallel Programming*, PPoPP '16, pages 39:1–39:2, New York, NY, USA, 2016. ACM.

[50] Timothy G Mattson, Beverly A Sanders, and Berna Massingill. *Patterns for parallel programming*. Addison-Wesley, Boston, 2005.

[51] U. Meyer and P. Sanders. -stepping: a parallelizable shortest path algorithm. *Journal of Algorithms*, 49(1):114–152, October 2003.

[52] William D. Nordhaus. The Progress of Computing. SSRN Scholarly Paper ID 285168, Social Science Research Network, Rochester, NY, September 2001.

[53] Mark H. Overmars and Jan van Leeuwen. Maintenance of configurations in the plane. *Journal of Computer and System Sciences*, 23(2):166–204, October 1981.

[54] David A. Patterson, John L. Hennessy, and Peter J. Ashenden. *Computer organization and design the hardware/software interface*. Elsevier/Morgan Kaufmann, Amsterdam; Boston, 3rd rev. edition, 2007.

[55] Vijay Saraswat, George Almasi, Ganesh Bikshandi, Calin Cascaval, David Cunningham, David Grove, Sreedhar Kodali, Igor Peshansky, and Olivier Tardieu. The asynchronous partitioned global address space model. In *The First Workshop on Advances in Message Passing*, pages 1–8, 2010.

[56] N. Satish, M. Harris, and M. Garland. Designing efficient sorting algorithms for many-core GPUs. In *IEEE International Symposium on Parallel Distributed Processing, 2009. IPDPS 2009*, pages 1–10, May 2009.

[57] Robert Schreiber. A few bad ideas on the way to the triumph of parallel computing. *Journal of Parallel and Distributed Computing*, 74(7):2544–2547, July 2014.

[58] Shubhabrata Sengupta, Mark Harris, Yao Zhang, and John D. Owens. Scan Primitives for GPU Computing. In *Proceedings of the 22Nd ACM SIGGRAPH/EUROGRAPHICS Symposium on Graphics Hardware*, GH '07, pages 97–106, Aire-la-Ville, Switzerland, Switzerland, 2007. Eurographics Association.

[59] J. A. Sethian. A fast marching level set method for monotonically advancing fronts. *Proceedings of the National Academy of Sciences*, 93(4):1591–1595, February 1996.

[60] J. A. Sethian. Level set methods: An act of violence - evolving interfaces in geometry, fluid mechanics, computer vision and materials sciences. *American Scientist*, pages 254–263, 1996.

[61] James Albert Sethian. *Level set methods and fast marching methods: evolving interfaces in computational geometry, fluid mechanics, computer vision, and materials science.* Cambridge University Press, Cambridge, U.K.; New York, 1999.

[62] Jeremy Shopf, Joshua Barczak, Christopher Oat, and Natalya Tatarchuk. March of the Froblins: Simulation and Rendering Massive Crowds of Intelligent and Detailed Creatures on GPU. In *ACM SIGGRAPH 2008 Games*, SIGGRAPH '08, pages 52–101, New York, NY, USA, 2008. ACM.

[63] Julian Shun and Guy E. Blelloch. Ligra: A Lightweight Graph Processing Framework for Shared Memory. In *Proceedings of the 18th ACM SIGPLAN Symposium on Principles and Practice of Parallel Programming*, PPoPP '13, pages 135–146, New York, NY, USA, 2013. ACM.

[64] Julian Shun, Guy E. Blelloch, Jeremy T. Fineman, Phillip B. Gibbons, Aapo Kyrola, Harsha Vardhan Simhadri, and Kanat Tangwongsan. Brief Announcement: The Problem Based Benchmark Suite. In *Proceedings of the Twenty-fourth Annual ACM Symposium on Parallelism in Algorithms and Architectures*, SPAA '12, pages 68–70, New York, NY, USA, 2012. ACM.

[65] Oliver Sinnen. *Task Scheduling for Parallel Systems.* Wiley-Interscience, Hoboken, N.J., 2007.

[66] David B. Skillicorn and Domenico Talia. Models and Languages for Parallel Computation. *ACM Comput. Surv.*, 30(2):123–169, June 1998.

[67] Yan Solihin. *Fundamentals of Parallel Multicore Architecture.* Chapman & Hall/CRC Computational Science. CRC Press, 2015.

[68] Daniel J. Sorin, Mark D. Hill, and David A. Wood. *A primer on memory consistency and cache coherence.* Morgan & Claypool, San Rafael, Calif., 2011.

[69] Juha Sorva. Notional Machines and Introductory Programming Education. *Trans. Comput. Educ.*, 13(2):8:1–8:31, July 2013.

[70] Ayal Stein, Eran Geva, and Jihad El-Sana. CudaHull: Fast parallel 3d convex hull on the GPU. *Computers & Graphics*, 36(4):265–271, June 2012.

[71] Robert Stewart and Jeremy Singer. Comparing fork/join and MapReduce. Technical report, Technical Report HW-MACS-TR-0096, Department of Computer Science, Heriot-Watt University, 2012.

[72] Rajeev Thakur and William D. Gropp. Improving the Performance of Collective Operations in MPICH. In Jack Dongarra, Domenico Laforenza, and Salvatore Orlando, editors, *Recent Advances in Parallel Virtual Machine and Message Passing Interface*, number 2840 in Lecture Notes in Computer Science, pages 257–267. Springer Berlin Heidelberg, September 2003.

[73] Stanley Tzeng and John D. Owens. Finding Convex Hulls Using Quickhull on the GPU. *arXiv:1201.2936 [cs]*, January 2012. arXiv: 1201.2936.

[74] Barry Wilkinson and C. Michael Allen. *Parallel programming: techniques and applications using networked workstations and parallel computers*. Prentice Hall, Upper Saddle River, N.J., 1999.

[75] Jianming Yang and Frederick Stern. A highly scalable massively parallel fast marching method for the Eikonal equation. *arXiv:1502.07303 [physics]*, February 2015. arXiv: 1502.07303.

[76] L. Yavits, A. Morad, and R. Ginosar. The effect of communication and synchronization on Amdahls law in multicore systems. *Parallel Computing*, 40(1):1–16, January 2014.

[77] Hongkai Zhao. A fast sweeping method for Eikonal equations. *Mathematics of Computation*, 74(250):603–627, 2005.

[78] Hongkai Zhao. Parallel Implementations of the Fast Sweeping Method. *Journal of Computational Mathematics*, 25(4):421–429, July 2007.

[79] Weizhong Zhao, Huifang Ma, and Qing He. Parallel K-Means Clustering Based on MapReduce. In Martin Gilje Jaatun, Gansen Zhao, and Chunming Rong, editors, *Cloud Computing*, number 5931 in Lecture Notes in Computer Science, pages 674–679. Springer Berlin Heidelberg, December 2009.

Index

Printed and bound by CPI Group (UK) Ltd, Croydon, CR0 4YY

23/10/2024

01777692-0008